Oh, Say, Can You See?

BORDERLINES

Oh, Say, Can You See?

The Semiotics of the Military in Hawai'i

KATHY E. FERGUSON AND PHYLLIS TURNBULL

BORDERLINES, VOLUME 10

 University of Minnesota Press

Minneapolis

London

The University of Minnesota Press gratefully acknowledges permission to reprint the following. Chapter 4 is adapted from the article "Narratives of History, Nature, and Death at the National Memorial Cemetery of the Pacific," by Kathy E. Ferguson and Phyllis Turnbull, in *Frontiers: A Journal of Womens Studies*, vol. 16, no. 2/3 (1996): 1–23. Chapter 5 is adapted from the article "Remembering Pearl Harbor: The Semiotics of the *Arizona* Memorial," by Phyllis Turnbull, in *Challenging Boundaries*, edited by Michael J. Shapiro and Hayward R. Alker, copyright 1996 University of Minnesota Press.

Published by the University of Minnesota Press
111 Third Avenue South, Suite 290
Minneapolis, MN 55401-2520
http://www.upress.umn.edu

Library of Congress Cataloging-in-Publication Data

Ferguson, Kathy E.
    Oh, say, can you see? : the semiotics of the military in Hawaii / Kathy E. Ferguson and Phyllis Turnbull.
        p.   cm.
    ISBN 0-8166-2978-1 (hardcover : alk. paper). — ISBN 0-8166-2979-X (pbk. : alk. paper)
        1. United States—Armed Forces—Hawaii.   2. United States—Armed Forces—Military life.   3. Fort DeRussy (Hawaii)   4. Hawaii—Social life and customs.   5. Sociology, Military—Hawaii.   6. Semiotics—Hawaii. I. Turnbull, Phyllis.   II. Title.
    UA26.H38F47   1998
    355'.009969—dc21   98-26195

Printed in the United States of America on acid-free paper

The University of Minnesota is an equal-opportunity educator and employer.

10 09 08 07 06 05 04 03 02 01 00 99   10 9 8 7 6 5 4 3 2 1

This book is dedicated to the memories of two of the Logsdon girls, my mother and her sister Moneta, and to that Arrowsmith girl, Edwina, my grandmother. PHYLLIS TURNBULL

And also to the memory of David Esquinazi (1930–1996) and Anita Lowry (1950–1996). KATHY E. FERGUSON

Contents

Acknowledgments

There are several things we ought to get out of the way at the start. We know we are transgressing: we know we are women writing about the military and we are Caucasian immigrants from the mainland writing about Hawai'i. We hope that our transgressive analyses will provoke and support questions raised about what we see as a militarized culture. We also hope that we have learned enough from those who are from here about how to live here without reproducing the colonial practices of those who preceded us.

We came to this topic from quite different ways and times, converging a few years ago to begin the coauthoring that has produced this manuscript. One of us, coming in 1954, took the dense military presence for granted until the Vietnam War dramatically decoded it. The other, arriving in 1985, and having already published a book on bureaucracy because she noticed it was all around her, noticed the same about the military here. And so . . . finally, because we teach in the same department at the University of Hawai'i, out of shared conversations a coauthoring demiurge was born.

As women, we are entering decidedly male territory, trespassing in ways that may elicit dismissal or contempt. Militaries often define themselves in opposition to women, revealing an anxiety about controlling those elements of the world or of themselves that are coded feminine. We invite those resistant elements to continue to analyze the military, to offer critiques and alternatives. We offer feminist analytic energies, gendered metaphors, and semiotic reading practices that are useful for unraveling the hyperpatriarchal character of military practices. The same (unruly) elements are often analyzing (and resisting) other historical conditions of a century-long colonial period, heavily coded white. Colonial legacies haunt contemporary identities, shaping the repertoires of available subject positions around the complex histories of racial and gender hierarchies, class privilege, and state power. We pursue conversations with a variety of political voices in Hawai'i, including Native Hawaiian activists, environmentalists, peace activists, feminists, and others. While we certainly do *not* speak for any of these groups, we join *with* them, hoping to amplify the availability of anticolonial, anticapitalist, feminist perspectives on what is called our state's second largest industry.

Our work has been much enhanced by our early collaboration with Mehmed Ali, whose vigilant archival work was matched only by his sharp eye for the practices of power within his own military unit. Joelle Mulford also assisted in constructing the earliest versions of this manuscript by serving as researcher. Both were birds of passage and have now left Hawai'i. Many others who live here have tried to nudge us or, when necessary, give us a shove to pay more attention to what we were saying or doing. They read drafts, offered gold mines of information, stories, histories, and sources. They tried to correct our spelling, dates, and facts. They did their best, but we take the heat for our mistakes. Those cherished people include Ho'oipo DeCambra, Roger Furrer, George Harrison, Kathie Kane, Louise Kubo, Nahua Patrinos, Noenoe Silva, and Ida Yoshinaga.

Other people whose assistance was generously and deftly provided include Scott Daniels, Carolyn DiPalma, the Hawai'i State Archives and staff, Nancy Morris, Karen Peacock and Chieko Tachihata at the Hawaiian and Pacific Collections at Hamilton Library, University of Hawai'i, Katharina Heyer, Arnold Hiura, Carol Langner, Paul Larned, Dale Madden, Norman Meller, Judy Rohrer, David Shapiro, Combat Veteran Murray Turnbull (Sgt., United States Army Air Force), and Lynn Wilson.

We thank the people who gave us their time and their thoughts in interviews, including U.S. Representative Neil Abercrombie, Judy Bowman, George Harrison, University of Hawai'i Vice President Eugene Imai, U.S. Senator Daniel K. Inouye, and public affairs officers on various military bases. Students in our Learning Community on Women and Men, Peace and War, and graduate students in our feminist theory and Hawai'i politics seminars, necessarily gave us their ears while also plying us with questions and enlarging our views on subjects. The American Friends Service Committee/Hawai'i continues to be an exemplar of reflective community participation. Our work has been enabled by grants from several units within the University of Hawai'i: the Office of Women's Research, the Social Science Research Institute, and the office of Dean Richard Dubanoski, College of Social Sciences. Carrie Mullen, our editor at the University of Minnesota Press, gave us a lot of rope and endless encouragement. Louisa Castner was a patient and skillful copy editor.

Many of our intellectual debts are evident in the pages of our text. We are swimming in an intellectual current with a variety of others, hoping to contribute to a cultural shift in the self-understandings available to citizens of contemporary nation-states. We particularly thank, for their comments on earlier versions of these chapters and for their insights into some of our knottiest problems: Wendy Brown, Eloise Buker, William Connolly, Tom Dumm, Cynthia Enloe, Renee Heberle, Manfred Henningsen, Jonathan Goldberg-Hiller, Tim Kaufman-Osborne, S. Krishna, Kirstie McClure, Chris Robinson, Michael J. Shapiro, Christine Sylvester, and Geoffrey White.

Gili, Oren, and Ari Ashkenazi and Suzanne Tiapula and Ruy Tiapula de Alencar have patiently supported the labor of this project; their sustaining energy is much appreciated.

Introduction

Everywhere you look in Hawai'i, you see the military. Yet in daily life relatively few people in Hawai'i actually see the military at all. It is hidden in plain sight.[1] This paradox of visibility and invisibility, of the available and the hidden, marks the terrain investigated in this book.[2]

For something to be in plain sight it must mark a variety of spaces, projecting itself into a number of landscapes. For something to be hidden it must be indiscernible, camouflaged, inconspicuously folded into the fabric of daily life. The key to this incompatibility is a series of narratives of naturalization and reassurance. The narratives of naturalization imbricate military institutions and discourses into daily life so that they become "just the way things are." The narratives of reassurance kick in with a more prescriptive tone, marking the military presence in Hawai'i as necessary, productive, heroic, desirable, good. The two narratives cross-fertilize and enliven one another—the military presence here is named as "natural," therefore as desirable, and constructive, therefore welcome. This discursive interbreeding between what is natural and what is good results in a tone of inevitability: what is, is good and in any case cannot be changed.

This book argues for four related claims about the military in Hawai'i. First, it is a particular kind of order, with identifiable consequences for Hawai'i's social, political, and geographic landscapes;

military order is not synonymous with order per se. Military order can become confused with Order in part by virtue of its migration into other institutions, such as schools, workplaces, media, and so on, so that a military model of authority and participation is reproduced. Alternative forms of order, such as those practiced by the indigenous people of Hawai'i, the descendants of contract laborers and other immigrant groups, feminists, and environmentalists, get marked as Disorder, as messy, inefficient, lacking competence.

Second, the military order is heavily written onto Hawai'i, marking literal and figurative spaces in manners both subtle and gross. On the more obvious level, there is the 16 to 25 percent of the island of O'ahu owned or controlled by the military (see chapter 1, note 2), the 16 percent of the population of the island who are military personnel or their dependents, the frequent presence of military planes and helicopters in the air, and military ships docked or offshore. On a more subtle level, names such as Fort Shafter, Schofield Barracks, and Pearl Harbor appear matter-of-factly as freeway exit signs and bus destinations; state license plates feature claims to veteran and wounded veteran status; military units make visible and well-publicized appearances in holiday parades; local businesses offer discounts to military personnel; troops returning from deployment receive extravagant media coverage. The connections between the extensive material presence of military installations in Hawaiian land and water and the indirect, seemingly casual markers of the military in everyday life are seldom articulated.

Third, and as a consequence of the first two factors, the military is thoroughly normalized within Hawai'i, sedimenting itself through accumulated familiarity into the everyday ways of life that produce what we experience as normal. Through its dense ordinariness it becomes constitutive of Hawai'i, not an interloper or even a visitor but a presence that simply and unproblematically belongs. The long and troubled history of conquest is muted by official accounts that fold Hawai'i neatly into the national destiny of the United States. Similarly, the relationships to places and peoples cultivated by Hawai'i's indigenous people and immigrant populations are displaced as serious ways of living and recalibrated as quaint forms of local color.

The particularity, ubiquity, and normalization of the military in Hawai'i all contribute to the discourses of naturalization by which the military is rendered unremarkable. When questions are nonetheless raised about the military presence, the discourses of naturalization give way to a different narrative register to respond to and deflect potential criticisms. These narratives of reassurance provide the needed legitimacy when dissonant perspectives are voiced, such as those offered by proponents of Native Hawaiian sovereignty, displaced farmers whose ancestral lands are inaccessible due to long-term military uses of the land, and environmental or feminist critics of big military budgets. It is our fourth claim that the narratives of reassurance mask a form of bribery, in which Hawai'i's economic and physical security can only be guaranteed by the military presence. Talk of base closings or economic conversion instantly raises complex fears about the loss of jobs and capital to the local economy as well as the military vulnerability of Hawai'i and the United States in a world perceived as threatening and insecure. The narratives of reassurance are tied to a set of economic and political claims that promise ruin and danger to those who might raise questions or pursue alternatives, and rewards to those who take the military presence as both natural and good.

These muted practices of naturalization and reassurance produce processes of militarization that are hidden in plain sight. Hawai'i as a site for these practices is a concentrated and intense example of a larger national/global process. By analyzing the particular expressions of militarization here, we hope to shed light on parallel practices elsewhere, and to link Hawai'i not only with American narratives of war and the state but also with other colonized islands in the Pacific. We aim to offer both a particular analysis of the military within a specific constellation of places called Hawai'i and a more general critique of American narratives of war and the state.

In "The Art of Telling the Truth," Michel Foucault (1988a) asks, "What is my present? What is the meaning of this present? And what am I doing when I speak of this present?" (89). Our methodological tool box has been selected for its ability to help us tell a history of Hawai'i's military present. We are concerned with how things come to be, and to be spoken: modes of representation, sites of memory, constructions of speech and silence; historical narratives of necessity and obligation, of sacrifice and security. We want to track these representational practices, looking at their construction, circulation, and interruption in Hawai'i's geographical and cultural spaces. A history of the present is a dissenting methodology in that it resists the familiar narratives and interrupts the ability to think that "the facts speak for themselves." We aim to tell a genealogy of some "facts," asking how and toward what end they are produced. We are concerned not with a complete or total fit between analysis and the world, but with an examination of how such "fits" come to be and whom they benefit. We are not trying to tell the one true story of Hawai'i's past, but to call into question the prior legitimacy claims of the dominant narratives and make room for others to appear. Our skepticism toward the dream of a common language, the big picture that brings everything together, does not leave us equally receptive toward all competing interpretations. Rather, we argue here for ways of thinking that illuminate the margins, contest the hegemonic stories, and enable stories more critical of established power to be told.

Our semiotic investigation of the marks and inscriptions of the military on Hawai'i lead us to listen attentively to the military's representations of itself, and to others' representations of the military. Our goals are to locate the particular representational elements that both conceal and reveal the military's presence and power; to illuminate the functions of these elements of representation as rituals of power; and to expand discursive space so that other voices can be heard. The practices of citizenship that are produced or foreclosed by the narratives of order and security are written onto Hawai'i with a military hand. Available understandings of citizenship, of what it can mean to belong to such an order, are embedded in the universe of possibilities that the dominant narratives produce. These include the kinds of claims that can be lodged against the order, the kinds of relationships that are appropriate, the varieties of responsibilities that can claim entitlement, the ways of being that are possible or legitimate. Militarized practices of citizenship that define and administer bodies—social, physical, and environmental—occlude alternative notions of citizenship and embodiment.

In her analysis of racialized normalizations of power in *Playing in the Dark*, Toni Morrison analyzes the energizing background presence of Africans in relation to white literature by way of the metaphor of the fishbowl:[3]

> It is as if I had been looking at a fishbowl—the glide and flick of the golden scales, the green tip, the bolt of white careening back from the

gills; the castles at the bottom, surrounded by pebbles and tiny, intricate fronds of green; the barely disturbed water, the flecks of waste and food, the tranquil bubbles traveling to the surface—and suddenly I saw the bowl, the structure that transparently and (invisibly) permits the ordered life it contains to exist in the larger world. (1992: 17)

We are following Morrison's lead in calling on our understandings of how arguments get made, how stories get told, "how language arrives" (17) to unravel the persistent productions and erasures performed by/in militarized spaces. Yet we depart from Morrison's metaphor (as so might she) in our thinking about fishbowls themselves. The image calls up a solid, heavy, finished container within which things move, while we want to sketch militarized frames of understanding that are both immensely pervasive and permanently incomplete. These frames work by consolidating complex processes into flat dichotomies and rigid hierarchies, by recruiting language and history into a one-size-fits-all account that is persistently hegemonic but never fully monolithic. We aim to listen for the silences, the necessary absences, the vigorous productions that parade as given. Our fishbowl is ruthless in its insistencies, yet it keeps springing leaks; its very ruthlessness toward a single pattern of insistences produces its leaks.

The particular fishbowl to which we are applying ourselves, the constellation of militarized practices that produces our present in Hawai'i and other sites of the national security state, is a touchy one. Hegemonies usually are. In the process of writing this book we have interviewed many people who shared with us a range of commitments, from those we might laud to those we abhor. Our goal is not to criticize or condemn the individuals who are or have been in the military, but to call attention to the devastating discursive and institutional consequences of militarized ways of being in the world. Our concern, ultimately, is with the constrictions that militarization places on democratic citizenship. We look around for the places in which citizens are created, and we see relentlessly militarized spaces, where only one model of what it means to participate in collective life is available. The model is hierarchical, authoritarian, bellicose. It is produced by a variety of interlinking hegemonies—capitalism, patriarchy, colonialism—that mark it metaphorically as male, white, western, affluent, and exclusive. We long for a different notion of citizenship—one that is suggested in many places but seldom vigorously enabled in contemporary Hawai'i or the United States. A more robust democratic citizenship entails critically thinking, persistently questioning authority, debating alternatives, respectfully engaging with a variety of othernesses, speaking and listening in a vigorous public space.

1

Traffic in Tropical Bodies

Hawai'i has the dubious distinction of being the most militarized state in the United States (Albertini et al. 1980: i). Estimates of the military's land holdings in the state, while often inexact, generally put the total at 5 to 10 percent.[1] By its own calculations, the military owns or controls 16 percent of the island of O'ahu (see fig. 1). Other sources put the amount at 23 percent (First Hawaiian Bank 1993b: 14). The best-known holdings include Pearl Harbor Naval Base (hosting the Naval Shipyard, the Submarine Base, the Naval Supply Base, and Westloch Naval Magazine), Hickam Air Force Base, Kāne'ohe Marine Corps Station, Schofield Barracks Military Reservation, Fort Shafter Military Reservation, Fort Ruger Military Reservation, Wheeler Air Force Base, Camp H. M. Smith (host to Commander in Chief, U.S. Pacific Command [CINCPAC], the command center of U.S. military forces from the Indian Ocean to California [United States Department of Defense 1990: appendix 1]. Major installations on other islands include Pacific Missile Range Facility at Barking Sands on Kaua'i; Pōhakuloa Training Area on the island of Hawai'i (locally known as the Big Island); and scattered holdings elsewhere. Even on the tiny privately owned island of Ni'ihau, where indigenous language and culture is preserved through isolation, there is a sporadic military presence (Adamski 1997: A-3).

Approximately 16 percent of the population of O'ahu consists of military personnel and their dependents.[2] The military is the second

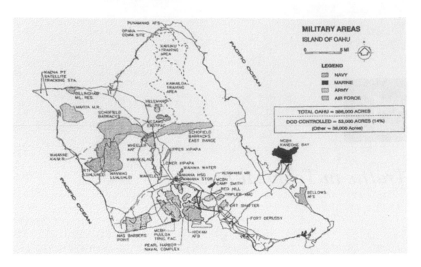

Figure 1. Military controlled land on the island of Oʻahu. From Hawaii Military Land Use Master Plan.

largest industry in the state, following tourism and preceding construction.[3] The Honolulu International Airport shares several runways with Hickam Air Force Base, treating incoming travelers to a preview of military power and presence. Military sites of memory, most notably the *Arizona* Memorial at Pearl Harbor and the National Memorial Cemetery of the Pacific, are must-see tourist destinations. The local newspapers are saturated with military coverage, ranging from reports on changes in personnel, to coverage of military "Good Samaritan" activities, to local protests against stray bombs and hazardous materials stored near schools. Three multilane freeways, one of which cuts through places sacred to some Native Hawaiians, connect military bases to one another. Reserve Officers' Training Corps (ROTC) and its junior version (JROTC) flourish in high schools and the university. Retired military personnel often settle in Hawaiʻi, pursuing second careers and/or sitting on the governing boards of a variety of local institutions. Military vehicles compete with commuter traffic, military bands march briskly in parades, military uniforms dot the human landscape, military names serve as freeway exit signs.

These observations constitute some facts about the military in Hawaiʻi. But the facts never speak for themselves; meaning does not dwell in objects but accrues through the narrative strategies by which the facts are recruited and made available for comprehension and contestation. "The most densely militarized state in the union" acquires this status not simply through the perceptible presence of military objects and events, but through the social and economic insinuation of the military into other institutions, and the cultural imbrication of military codes, symbols, and values into daily life. These latter processes flag not the military per se but militarization as a dynamic, contested process of constituting a particular kind of order, naturalizing and legitimating that order, while simultaneously undermining competing possibilities of order.

Telling a history of Hawai'i's militarized present entails attention to the process of *how* these observations are facts, how they qualify to enter into discourse, to be spoken and understood, contested or taken for granted.[4] This kind of history, Michel Foucault's genealogy, attends to "the concrete body of a development, with its moments of intensity, its lapses, its extended periods of feverish agitation, its fainting spells" (Foucault 1977: 145). Hawai'i's militarized present was produced and is maintained by a confluence, both interactive and discontinuous, of several streams of order. Most notable among these narrative productions are those by explorers, missionaries, sugar planters, soldiers, and tourists. Each stream carries a variety of intertwined accounts by the various newcomers about what Hawai'i offers and what it lacks, about how their projects can fulfill Hawai'i's promises, supplement its voids, and avoid its entrapments. These authoritative projects have always entailed the enforced movement of a variety of bodies across a variety of borders and the persistent transgression and frantic reinforcement of a range of critical boundaries. Numerous violences have attended these movements and markings: the material violence of displacement, uprooting, and resettlement; the discursive violence involved in reading a place through the lenses of one's own desires; the ontological violence of writing a particular kind of order onto bodies and spaces.

Reflecting on the militarization of the United States in the past fifty years, Michael Sherry refers to it as "the process by which war and national security became consuming anxieties and provided the memories, models, and metaphors that shaped broad areas of national life" (Sherry 1995: ix). Tracking this complex process in Hawai'i requires unraveling prior sets of anxieties and desires, since Hawai'i's militarized present is "merely the current [episode] in a series of sub-jugations," not the culmination of a singular historical logic but the friction and slippage, inertia and momentum of "the hazardous play of dominations" (Foucault 1977: 148).

Central to the streams of order converging and rebounding on Hawai'i's militarized present are particular organizations of sex, race, and class as essential vectors of power. Sex/gender, race/ethnicity, and class/property energize, consolidate, and disperse the prevailing play of dominations in both the semiotic and the material domains of social life.[5] As Anne McClintock powerfully demonstrates, these three critical dimensions of power relations in colonized places always act as "articulated categories" that "come into existence *in and through* relations to each other" (McClintock 1995: 5). Each is always already marked with the historical patterns and practices, the cultural logic and ambiguity, the institutional distributions and erasures carried by the others. In such contexts, race/ethnicity always affixes itself to laboring or nonlaboring bodies and to gendered relationships; sex/gender always marks persons to whom particular colors and classes are simultaneously attributed and enforced; labor is always organized, and property defined and distributed, among groups also ordered around reproductive functions, sexual practices, and color codings. Imperial conquest is interdigitated with the cult of domesticity and the global political economy (McClintock 1995: 17). The well-bred classes carry the white man's burden and speak in the voice of the father. Women's status as property feeds Europe's self-ordained cultural destiny in the "empty lands" and its historical mandate to industrial progress. Male power, white power, and commodity capital; female sexuality, cannibalism, and plunder; child-rearing practices, missionary schools, and plantation labor; population control, origin stories, and the market; penetration/emasculation, conquest/engulfment, appropriation/absorption. The terms of this energetic, interactive triangle chase and dodge around one another, powerfully enabling each other while sometimes interrupting or confounding the patterns their dances produce.

These three interanimating, mutating axes of power operate on both the material and semiotic levels, requiring simultaneous attention to the tangible productions and distributions of land, labor, schools, churches, families, and wars, and to the acts of speech and silence that produce and enforce meaning claims in discourse. Discourse does not relate to the material realm as a hidden meaning standing behind the surface, but as an "unsublatable dialectic of saying and seeing" encountering the persistent opacity of objects as well as their availability (Jay 1994: 398). "This discourse," Foucault comments in a different context, "forms a tissue where the texture of the verbal is already crossed with the chain of the visible" (quoted in Jay 1994: 398). What can be said/written/understood is always already interacting with what can be seen/grasped/seized in ways that are both mutually constitutive and condemned to incompleteness.

The tension between availability and resistance to the interaction of discourse and material lends itself to—perhaps even gives rise to—the uneasy combination of fears and longings in the colonial encounter with Hawai'i. The prevailing emotional registers in which colonial discursive and institutional practices are most commonly intoned reflect a restless mixture of desire and anxiety. In McClintock's words,

> the inaugural scene of discovery becomes a scene of ambivalence, suspended between an imperial megalomania, with its fantasy of unstoppable rapine, and a contradictory fear of engulfment, with its fantasy of dismemberment and emasculation. The scene, like many imperial scenes, is a document both of paranoia and of megalomania. (McClintock 1995: 26–27)

The gaze of various newcomers onto the Hawai'i they encounter combines, in differing proportions, the three terms of the gaze that Roland Barthes attributes to science:

> in terms of information (the gaze informs), in terms of relation (gazes are exchanged), in terms of possession (by the gaze, I touch, I attain, I seize, I am seized): three functions: optical, linguistic, haptic. But the gaze seeks: something, someone. It is an anxious sign: singular dynamics for a sign; its power overflows it. (Quoted in Jay 1994: 441)

Hawai'i's colonizers brought with them both a profound sense of entitlement and an abiding fear of engulfment. While the information they gathered and the relations into which they entered were largely guided by the seizures they sought, there remained an unsettling incompleteness both animating and disturbing to the colonizers. Hawai'i's perceived deficiencies provoked both desire (take it, fill it, make it ours) and anxiety (it's different, it's not like us, it's looking back at us). Like the vagina (dentata) that is thought both to require the penis for fulfillment and simultaneously to threaten to sever it, Hawai'i both beckons and disturbs its newcomers.

The beginnings of Hawai'i's colonization lie in its encounters with western explorers, missionaries, entrepreneurs, and sugar planters, all propelled by different mixtures of longing and trepidation. Each group actively engaged the available geographical and cultural horizons for opportunities to establish "the spatial forms and fantasies through which [their] culture declares its presence" (Carter 1987: xxii). The explorers encountered a place they defined as largely empty of meaning, lacking in culture, but available for western expansion. The missionaries found a people they defined as dark, mysterious, lacking civilization but capable of being domesticated. Entrepreneurs and sugar planters found the people lacking industry, the land uncultivated but a promising venue for profit once an appropriate labor force could be secured. The military saw/sees Hawai'i as strategically important and in need of defense, which imported American soldiers can supply. The traffic in workers and soldiers finds parallel in "the world's peace industry," the commerce of bodies across borders that tourism produces and celebrates (Louis D'Amore quoted in Diller and Scofidio 1994: 24).

Insinuated thoroughly into the political, economic, and cultural practices of colonialism, national security, and international tourism are the ubiquitous markings of gender. Western intrusions into Hawai'i—from early explorers, traders, and missionaries, to planters, diplomats, and military leaders, to travel agents, airline companies, and foreign investors—have seen Hawai'i as a welcoming feminine place, waiting with open arms to embrace those who come to penetrate, protect, mold, and develop, while simultaneously lacking that which would make it fully realized (and which the intruders conveniently believe themselves to possess). Maps of Hawai'i from Captain James Cook's expeditions represent Hawai'i with soft, curved, breast-like mountains and mysterious coves and bays (David 1988: xl, 136). Missionary accounts of "the natives" emphasize their darkness, their eroticism, and their lack: dark-skinned, living in darkness; naked, unashamed, promiscuous; lacking in writing, lacking in modesty, lacking in the knowledge of God. Planters found a different lack—lack of industriousness, frugality, punctuality, property, devotion to accumulation. Military planners found a lack of awareness of foreign enemies, an absence of appropriate defense technology, a properly mascu-

Figure 2. The fleet's in! Courtesy Hawai'i State Archives.

line political hierarchy wanting (fig. 2). Tourism's representational economies figure Hawai'i's femininity in its more positive vein—a soft, smiling, tender place that waits to be shared, enjoyed, consumed. Both versions of femininity—as lack and as availability—configure the representers as entitled to what they see. Hawai'i as inadequate female or welcoming woman is still defined by the intruder's gaze.

THE FIRST VOYAGERS

Polynesian voyagers making first human landfall on one of the is-lands[6] could have described their landing place as empty and rela-tively inhospitable; barren, rock-strewn, and lacking edible plants and animals, it was starkly different from the Marquesas Islands from which they sailed. The cluster of volcanic islands was remotely located in the Pacific; any vegetation and animal life found there had drifted as spores or seeds in the atmosphere and as passengers on rafts of floating debris.[7] Yet the new home of the early voyagers of-fered a space in which their labor and cultural practices could create a fertile and sustaining land. How many voyagers came, how many might have been lost at sea, and why they undertook the long dan-gerous voyage are questions without answers. Perhaps they were fishermen blown off course; perhaps they were banished; perhaps observing the annual migrations of plovers and a few other birds, they were fired by the possibilities of other wheres; whatever the provocation, their gaze sought something.

It is likely that they were able to return to their home and that others followed them in narrow outrigger voyaging canoes, bringing with them virtually all of the cultigens that soon formed their rich, nourishing agricultural base.[8] They brought Hawai'i into existence with their colonizing labors; lacking metal, clay, and animal labor, they transformed the land through ingenious use of trees, fibers, stone adzes, and staves. With rock and stones, they lined irrigation flumes, built roads and paths, constructed temple platforms; from trees they fashioned their familiar sailing canoes (Bushnell 1993: 11).

A second wave of voyagers arrived from Tahiti around the four-teenth century. Linguistically and culturally related to the first Poly-nesian voyagers, this group took over as invaders in some respects. They installed new *kapus* (prescribed rules for relationships between persons and gods, each other, and things around them), which in-creased the distance between the chiefs and the people. New rituals of domination, most notably human sacrifice, increased the hier-archic structure of the society (Bushnell 1993: 11–12). Despite such changes, however, most people prospered. They were healthy and well fed (except when natural disasters wiped out their crops). They could read, feel, and hear the earth for its signs, recognizing wind and ocean patterns, differentiating among innumerable types of stones, rocks, insects, plants. They composed and remembered epic poems of several thousand lines. For the world they had called into existence, they had ten different words meaning beautiful (Bushnell 1993: 14). After several thousand years of settlement and the creation of Hawai'i, the colonizers had become the indigenous people.

WESTERN DISCOVERERS

Subsequent voyagers were more discoverers than explorers, to use Paul Carter's distinction. Discovery rests on "the assumption of a world of facts waiting to be found, collected and classified, a world in which the neutral observer is not implicated," while exploration "lays stress on the observer's active engagement with his environment" (Carter 1987: 25). Both are caught in the three functions of the gaze that Barthes attributes to science—informing, relating, possessing—although in somewhat different ways.

The "discovery" of Hawai'i was an epiphenomenon of a voyage taken by Captain James Cook, on behalf of Britain's Admiralty Lords, to attempt to find "a Northern passage by Sea from the Pacific to the Atlantic Ocean" (Cook 1967, 3: ccxx).[9] (Their possessive gaze not only assumed the existence of a passage, and their entitlement to access, but that it would be navigable as well.) Cook landed briefly on the northwest island of Kaua'i in January 1778 and put ashore a small watering party of men—the first tourists—who, returning the next day with water, left behind the microbes of venereal diseases, an exchange of fluids presaging subsequent disasters for the Native Hawaiians. Cook's voyages carried the imperial gaze and command in the instructions he was given, which always included, as a matter of course, the creation and maintenance of boundaries. He was directed to take possession of "convenient Situations in the Country in the name of the King," leaving marks and inscriptions as "First Discoverers & Possessors" (Cook 1967, 2: clxviii).[10] At this ceremony, the colors were flown and the men usually fired off a volley. Beyond these manly imperial tasks, Cook was also instructed to spy out the land; survey and make charts of harbors and headlands; collect specimens of minerals, valuable stones, and seeds, shrubs, and grains; "observe the Genius, Temper, Disposition and Number of the Natives or Inhabitants" (Cook 1967, 2: clxviii). The scientific gaze was fully satisfied in two of Cook's voyages[11] as his ships carried botanists and artists to create the archive of "the world of facts waiting to be found, collected and classified" (Carter 1987: 25). Cook himself was selected for the voyages by virtue of his successful participation in the gaze that informs and the gaze that relates: he was noted for his meticulous draftsmanship and his brilliant navigating. By contrast to the later American explorations, Cook's voyages of discovery were the Enlightenment afloat.

Despite Paul Carter's sensitive interpretation of Cook as an explorer rather than discoverer, Cook's voyages nevertheless fed easily into the third gaze, the imperial project of reducing the world's variety into uniformity, universal knowledge, and possession (Carter 1987: 16). The voyages contributed to the ontological violence of writing a new imperial order onto bodies and spaces. His ship was a metaphor for that order, reproducing the social structure of England in its separation of forecastle and quarterdeck, and manned by a more disciplined crew than the men who sailed with the buccaneers and desperadoes of earlier centuries.[12] His technologies reflected Enlightenment improvements: Cook knew where he was and was able to keep the ship on course with sextant and chronometer. He had technologies unavailable to earlier sailors: time and longitude began at Greenwich.[13] Cook's sailors were somewhat healthier than earlier ones as a consequence of available medical knowledge. His crews did not succumb to the earlier occupational disease of scurvy, for antiscorbutics were carried as ships' stores and were often brewed of the flora of new places.

The tension between desire and anxiety in the explorer's gaze is evidenced in Cook's mandate to cultivate friendship with the natives while being on guard against surprises. With the natives' consent, he was to take possession of their land. This unlikely combination of anticipated events becomes intelligible in light of the uneasy interaction of familiarity and strangeness in the colonial project. The new world was strange, different, other; yet the guidelines about how to act in this new world reflect some very familiar understandings of gender, race, and class. In McClintock's words:

> Knowledge of the unknown world was mapped as a metaphysics of gender violence—not as the expanded recognition of cultural difference—and was validated by the new Enlightenment logic of private property and possessive individualism. In these fantasies, the world is feminised and spatially spread for male exploration, then reassembled and deployed in the interests of massive imperial power. . . . In the

minds of these men, the imperial conquest of the globe found both its shaping figure and its political sanction in the prior subordination of women as a category of nature. (1995: 23–24)

McClintock emphasizes the liminality of the explorers' condition, "on the margins between known and unknown," where they "became creatures of transition and threshold" (1995: 25). Yet the new dangers they faced are represented in the most familiar of terms: dark, unpredictable, primitive, feminine others whose friendship might be cultivated at the same time that their land is stolen. The explorers have in some ways left the old order behind since, for example, they would probably not have thought it possible to cultivate the friendship of a fellow Englishman while seizing his property. Yet the definition of whose land was available for such seizure, and whose sentiments naive enough to permit it, reflects ready imperial racial classifications.

On his return from the northwest coast of New Albion ten months later, Cook put in to a bay on the lee side of the island of Hawai'i to refresh his men, to repair the ships, to take on new stores of food and water. As a matter of course, he established an observatory, a site for the development of the observing gaze. The liminal space of the shore was made even more ambiguous in that Cook and his ships had arrived at a time and from a direction that coincided with the Hawaiian observance of the return of the peaceful god *Lono* from *Kahiki* (the ancestral/sacred land) to displace the war god *Kū*. Of such a collision between two meaning systems, Marshall Sahlins has argued that people engage one another within their own cultural categories for making sense of events. In this way, Cook was treated as *Lono* on Hawai'i in the latter part of 1788 (Sahlins 1995: 212). His party was fed by levies raised by the high chief; by the time of Cook's departure several months later, which coincided with the season for *Lono* to return to *Kahiki,* the area around the bay was depleted of food. After a mast broke a week or so later, he returned to the bay to repair it and encountered unexpected (to him) hostility from the chiefs. No longer taken to be *Lono,* he was received differently. Some of his ship's stores and implements were taken (not given or bargained for), watering parties were harassed, and, finally, the ship's cutter was seized; race and class were quickly, distinctively activated in a clear-cut dispute over boundary and property.

In order to regain the cutter, Cook attempted to entice the high chief to his ship, planning to hold him there until the ship's boat was returned, but the chief refused to go. Simultaneously, the word came to the Hawaiians that a different chief had been killed by one of Cook's troops at the far end of the bay. In the melee that followed, Cook was killed by the Hawaiians as he was walking back toward his small boat protected by armed marines. Cook's men escaped to their ships leaving their dead captain behind. Cook's body was removed by Hawaiians and given the treatment due a high chief.[14] At the end of that ritual period, some of the stripped bones of Cook were returned to the English; a *kapu* was laid on the bay to permit Cook's body a conventional seaman's burial, and peace was restored. Cook's ships left Hawaiian waters for good shortly afterward. The Enlightenment tools that allowed Cook to read the sea accurately, and to organize his ships effectively, were of little use in reading the cultural practices of the Native Hawaiians. While Cook's gaze was less haptic than that of subsequent plunderers and opportunists, it did not occur to him to ask how the Hawaiians looked at *him*. An "expanded recognition of cultural difference" might have allowed Cook to predict the Hawaiians' hostility to his unwelcome return, to locate their lavish earlier treatment as well as subsequent rejection in an indigenous political economy of sacredness and sharing.

A few years later, Captain George Vancouver arrived, in command of another official British expedition to the Pacific. One of his goals was to complete the exploration of the North American coast for the passage still lusted after. In the course of his voyages in the 1790s, Vancouver spent several months in Hawai'i at different times. He came to know most of the chiefs and cultivated a particularly strong relationship with Kamehameha, the high chief of the island of Hawai'i. Kamehameha later conquered the other islands and became king of Hawai'i. Vancouver saw himself as a statesman/adviser to Kamehameha and others of them. Always with an eye toward accession of the islands by Great Britain, he warned Kamehameha against the dangers of letting foreigners (other than himself) stay in the islands. Other domesticating movements included the introduction of cattle, sheep, goats, and geese to the islands. In yet another gesture of rewriting an order on the land and people, he had his carpenters assist in building a warship for Kamehameha, although the promised cannons were not delivered for several decades. Under Vancouver's direction, Kamehameha's men were taught how to drill as a body of soldiers (Kuykendall 1957: 37–44).

This necessarily brief account of early encounters with Hawai'i may risk reducing Hawai'i to a static stage on which imperial history is enacted. We want to avoid the kind of historical narrative that Paul Carter has called "diorama history—history where the past has been settled even more effectively than the country" (Carter 1987: xx). "Hawai'i" is not a preconstituted place on which destiny unfolds. The streams of order converging on Hawai'i were neither unanimous nor simple. Military, commercial, and exploratory contacts were intermingled in varying ways and times. Diorama history takes for granted "that the newcomers travelled and settled a land *which was already there*" (Carter, 1987: xxi, emphasis in original). We aim at a more dynamic history, in which each gaze on Hawai'i is seen as a constitutive gaze, an active expression of desire that brings particular Hawai'is into existence as cultural/political spaces.

Not all gazes are the same. Where Cook's voyages of discovery more heavily emphasized the gathering of the facts waiting to be found about the world, the United States Exploring Expedition of 1838–1842, commanded by Navy Lieutenant Charles Wilkes, was distinctly more commercial and martial in its goal, with scientists playing an auxiliary role (Herman 1995: 68). "Fitted out by national munificence for scientific objects," the expedition of two war sloops, a brig, a store ship, and two tenders was to explore and survey the Pacific, to determine the existence of "all doubtful islands and shoals, as well as to discover and accurately fix the position of those which lie in or near the track of our vessels in that quarter, and may have escaped the observation of scientific navigators" (Herman 1995: 68). The inadequacy of the informative gaze was, in the eyes of the American Navy, that it was only weakly possessive. The Wilkes expedition was to remedy this deficiency, to "sweep the broadest expanse of the ocean" possible "in those latitudes where discoveries may reasonably be anticipated" and to search out harbors where American whalers and other commercial vessels might shelter. In this respect, the intention of the government to keep a squadron of the fleet cruising in the Pacific in the future was announced (Herman 1995: 68). Despite no official instructions concerning Hawai'i, the Wilkes expedition spent more time in Hawai'i than any other place during the voyage (Sprague 1988: 113). Wilkes's crew spent nearly seven months exploring the four main islands of Hawai'i, acting as advisers to King Kamehameha III on legal matters, establishing observatories, and continuing the imperial mode.

Wilkes's ascension of Mauna Loa, a volcano on the island of Hawai'i, was made on the backs of a few hundred native Hawaiians, four of whom carried Wilkes in a sedan chair at times.[15] In order to effect discipline and order among the natives, each was designated by a tin disk, "painted of different colours, so that the wearers might be known to us and mustered without having recourse to their names or asking any questions" (Wilkes 1845, 4: 115). They were accompanied as well by a

> large number of hangers-on, in the shape of mothers, wives, and children, equalling in number the bearers, all grumbling and complaining of their loads; so that wherever and whenever we stopped, confusion and noise ensued. I felt happy in not understanding the language and of course was deaf to their complaints. (Wilkes 1845, 4: 118)

While Wilkes, unlike the Spanish conquistadors two centuries earlier, at least recognized that the natives spoke a language, he had no intention of entering into it, not even to learn his workers' names (Todorov 1984: 29–34). Soon after,

> while we were getting a slight nap, Dr. Judd was engaged in superintending the distribution of food to the multitude, during which time much confusion and noise existed. The natives put me in mind of wild beasts in this respect; they seldom make any noise unless their appetite and ease are in some way concerned. (Wilkes 1845, 4: 119)

Further along the ascension, and apparently after Wilkes fired a mortar at daybreak

> a rebellion was found to exist among the natives in the camp, that threatened to upset all our plans . . . we were obliged to defer our departure. Dr. Judd soon detected the ringleaders, one in particular who was holding forth to the Kanakas, advising them, as they now had me in their power, to strike for higher wages; for, if they did so, we should be obliged to pay them double, or anything extra they might ask for. He was at once made an example of by being turned out of the camp and sent away. . . . This had the desired effect, and the rest signified their willingness to go forward. . . . From this I well knew that no confidence was to be placed in the natives. (1845, 4: 125)

The natives had more shortcomings, as Wilkes was to discover. "The practice of medicine was not known in ancient times in Hawaii" (1845, 4: 47), he intoned (about a society in which healers were held in high regard); they "appear to have but little knowledge of astronomy" (1845, 4: 41) (while their long-distance navigational practices succeeded centuries earlier); moreover, "the natives . . . generally show very little attachment to their children" (while the extended family system placed very high value on children); and, he briefly concluded, "I should not be inclined to believe there is much natural affection among them; nor is there apparently any domestic happiness" (1845, 4: 45). At the same time he calculated that the natives would never overcome their natural laziness and sloth; "a native's idea of luxury does not extend beyond poe [sic] and fish, with which he usually seemed satisfied, and when they are obtained ceases all exertion" (1845, 4: 65). Finding the natives wanting in both ambition and greed, Wilkes seemed affronted by the absence of the desire-anxiety coalition and eager to export it. A man ahead of his time, he concluded that it would be necessary to make them feel "artificial wants, which cause them to look for employment" (1845, 4: 220). Wilkes identified lands that could be cultivated with sugarcane, cotton, wheat, and other American crops and located streams that might be used for agricultural irrigation and mill seats (1845, 4: 61, 65, 70, 79, 92). Finally, the intrepid explorer visited the area now known as Pearl Harbor and after an extensive survey anticipated an exciting penetration of the land, one that could accommodate many larger imperial members:

> The depth of water at its mouth was found to be only fifteen feet; but after passing this coral bar, which is four hundred feet wide, the depth of water becomes ample for large ships, and the basin is sufficiently extensive to accommodate any number of vessels. If the water upon the bar should be deepened, which I doubt not can be effected, it would afford the best and most capacious harbour in the Pacific. As yet there is no necessity for such an operation. (Wilkes 1845, 4: 79)

Moving quickly from the claims of racial idleness to the myth of empty lands, Wilkes performed a founding imperial gesture: the indigenous people are both inadequate and absent, not very good and not there at all. This gesture enacts a gendered racial dispossession—

it establishes the people as lacking crucial human attributes, and the land as available because no one is using it properly.

MISSIONARIES

Where the discoverers, as well as subsequent planters and other entrepreneurs, were to see an empty or virginal land, one capable of great fecundity, the New England missionaries foresaw a space filled with persons who for "long and dismal ages of darkness" had been "perishing for lack of knowledge," a population on whom the "Sun of Righteousness" had never risen, and who were living in the "rudest state of uncultured man."[16] For the seventy-three missionaries who, in three decades, constituted The Mission, the wish was to become father to the fact. As carriers of the colonial order, they violently elaborated the narrative overriding the indigenous social order with their own through an interplay of intractable maintenance of boundaries and boundary assaults. The class ironies of the missionaries' project—they came to do good and ended up doing well—have been frequently noted.[17] Less has been said about the male megalomania and the paranoia of the gendered order of The Mission itself.

For McClintock, the megalomania announces itself in the feminization of land, a strategy she terms a "violent containment" (1995: 23). The move by The Mission to heathenize the Hawaiians was similarly a manic act, also cohabited by paranoia. The darkness that the missionaries had pledged themselves to end was at the same time boundless and threatening, arousing the fear of loss of their own boundaries. To avoid their engulfment by the disorder of the unknown, they zealously rode shotgun on their own perimeters and organized mapping expeditions into the liminal space of darkness. Their efforts at neutralizing what McClintock terms the fears of "narcissistic disorder by reinscribing, as natural, an excess of gender hierarchy" (1995: 23) were written all over the institutional practices of the governing board, the American Board of Commissioners for Foreign Missions and were constitutive of the daily practices of missionary families in Hawai'i. Carefully defined orderings of bodies, sexuality, and families transformed threatening places into familiar domestic spaces. The strict resistance by the board to dispatching single men as missionaries was based, as Patricia Grimshaw writes, on "the experience of celibate men in the Tahitian mission of the London Missionary Society [which] had established clearly that, in the midst of a Polynesian community, celibate men were at risk from the sexual openness of the society" (1989: 6). Fearing the moral contamination and descent into sin threatened by sexual congress with native women, yet requiring the services of women who could be counted on to "[serve] discreetly at the elbow of power . . . upholding the boundaries of empire and bearing its sons and daughters," the board mandated that missionaries be married prior to their departure (McClintock 1995: 6). Facing this requirement, some missionaries approached the acquisition of a wife in much the same way as they must have done in supplying themselves with the "articles necessary as an outfit to the Sandwich Islands . . . a sufficient supply for three years" (Simpson 1993: 28). In place of buying brides, they interviewed for them; Grimshaw credits the board with actually brokering some of the marriages. Some indication of the excess of the gendered order is evidenced in the marriage of Dwight Baldwin and Charlotte Fowler, who sailed for Hawai'i one week after their first meeting (Grimshaw 1989: 12).

Once in Hawai'i, the patriarchal order unfolded, boundary patrols were established, a new technology of knowledge instituted, and domesticity urged on the indigenous people. The order quickly begat iself. Seventy of the total of seventy-six missionary wives who lived in Hawai'i for any length of time bore children, usually at regular intervals (Grimshaw 1989: 89). Missionary wife Sarah Lyman, a recent history of the family reveals, was evidently a bit out of step since it was thought remarkable that she "did not get pregnant for more than a year after the Lymans arrived in the Islands" (Simpson 1993: 62). Thirty-eight wives who lived in Hawai'i during their fertile years bore 250 infants (Grimshaw 1989: 89). In positioning them as helpmeets, as the board of commissioners did in its instructions (*The Promised Land* 1819: ix), the wives reproduced "the gender division of labor reminiscent of the domestic economy of many small business or professional households in New England" (Grimshaw 1989: 101). Their boundary maintenance work, conducted without the usual material resources of their New England homes, consisted of creating a comfortable home as a reassuring basis for the work of the male, and providing a "suitable" environment for the children of his name. Domestic practices within missionary homes indicate considerable border anxiety. *Dirty, stick,* and *kiss,* key words of a process of punishment and reward, were acquired early by missionary babies (Grimshaw 1989: 139). Toddlers and children in the language-acquiring stage increased the anxiety over boundaries, for even under "normal" circumstances the appearance of Satan in children's ways was always a threat. In these alien circumstances, security had to be heightened to forbid their children from learning the Hawaiian language, thought to be heathen and lewd, and to prevent their playing with the offspring, judged naturally depraved, of the people who lived there.[18]

Males of the mission mapped the liminal dark space, respatializing it as "posts" to which they were assigned, conducting Christian services, riding and walking around their "districts," and rigorously attempting to write their Christian narrative on the people. Some kept the boundaries clear for themselves through the periodic general meetings that were held, keeping their own notes on the general and particular practices and appearances of depravity visible to them, or through their correspondence with families back home, through their reports to the board of commissioners, or through their publishing activities. The males established an orthography for the Hawaiian language, imported a printing press, and went to work to transform an oral culture into a proper written one (see Buck 1993). Crucial here were illustrated primers, lexicons of the imperial and Christianized order (baffling as pictures of Noah's Ark, wolves, and Old Testament figures might have been to the would-be readers.)[19]

Hawaiian bodies were particularly threatening: comfortably large, half-clad in the eyes of the beholders, and bearing none of the defining marks of a familiar order as they went about their hedonistic and heathenish ways of life.[20] The hula in particular seems to have represented the greatest threat of engulfment and, conversely, male anxiety. It was not necessary to understand the words of the accompanying chant to interpret the energetic pelvic thrusts of the *mele ma'i*, but, obsessed with that movement, the missionaries could easily have missed (or, perhaps, perceived all too well) the energy and grace of the erotic. To keep this threat at bay, though never fully successfully, the missionaries instituted two kinds of restrictions: they clothed the offending bodies and discouraged hula, on the one hand, and rigorously restrained their own foreign bodies in long sleeves, high collars, cravats, trousers, long skirts, bonnets, and bound hair, on the other.

A final manic move in the process of mapping domesticity onto the social space of the Hawaiians was the attempt to introduce the concept of marriage and female submissiveness among those native to Hawai'i. Ultimately, the conjugal, autonomous, exogenous family order of The Mission was by itself no match for the dense kinship relations among Hawaiians, through whose extended families children were cared for and food was caught, grown, and pooled. Frequent visits among these rich social networks involved much travel about and between the islands by the natives, whose shifting about was interpreted as shiftlessness by their would-be tutors. Missionary efforts to carry out the conversion of the Hawaiians to the cult of domesticity suggest the interactive enabling of patriarchal domestic order and racist imperial order. Dark people are featured in colonial discourse as "gender deviants, the embodiments of prehistoric promiscuity and excess, their evolutionary belatedness evidenced by their 'feminine' lack of history, reason, and proper domestic arrangements" (McClintock 1995: 44). The hierarchical relations between women and men in European domestic space offered just the right model, in imperial eyes, for organizing relations between dark people and white men in colonial spaces.

By 1832, however, it had become apparent to the missionaries who conducted the first rough census that much of the traffic by native bodies had become one-way.[21] The number of the indigenous population had plummeted to 130,000.[22] Whether the number plunged from the 400,000 estimated by Cook's expedition or the 800,000 to 1,000,000 figure argued recently by David Stannard is, despite its significance otherwise, not at issue here (Stannard 1989). Our point is that the reordering of the land and people was more easily accomplished by the loss of those thousands of voices, and the rupture of the social relations and ways of life of the people of old. The Hawaiian bodies lacked not the proper order, Christian, mercantile, or literate, but, as implied by the subtitle of O. A. Bushnell's book, *Germs and Genocide,* the crucial antibodies against the invading bodies.

For the missionaries, it was clear that the licentious ways of the Hawaiians had caught up with them, thus confirming the rightness of the Mission's reordering goal. The Reverend Artemis Bishop, writing in *The Hawaiian Spectator* in 1833, attributed the near-extinction of the Hawaiians to licentiousness (of prostitution and veneral diseases) and their adoption of civilized vices such as alcohol (Bushnell 1993: 228). Fifty years later, the Reverend S. E. Bishop attributed it to the promiscuity of Hawaiian women, an ignorance of nursing, the Hawaiian "habitual indifference . . . to sanitation," *kahunas* (priests) and sorcery, idolatry, and "the wifeless Chinese" (Bishop 1888: 10, 15).[23] Proclaiming another leg of the argument, Walter F. Frear, one-time territorial governor of Hawai'i, opined that

> the advent of the white man in the Pacific was inevitable, and especially in Hawaii, by reason of its size, resources, and most important, its location at the crossroads of this vastest of oceans, rapidly coming into its own in fulfilment of prophecies that it was destined to become the chief theater of the world's future activities. (Frear 1935: 6)

Pursuing imperial history with a vengeance, Frear added that

> increasing knowledge . . . seems to show that most Pacific peoples were already on the toboggan before the whites came, that the latter merely greatly accentuated the decline, that the causes were largely psychological, that no uniform rule can be affirmed as to the various groups, and that there is no impossibility [*sic*] of effecting, through appropriate remedial measures, a mental and physical balance which will permit or insure survival or increase. (Frear 1935: 6–7)

While the gazes of discoverers and missionaries (as well as subsequent colonizers—the planters, the military, and promoters of tourism) strategically encompassed the rulers and chiefs, the common people were but motes in their eyes at best. Unindividuated, taken to lack personhood, the commoners were a distorted screen on which the carriers of the new orders projected their desires and rages. The dilemma inherent in the missionaries' gaze is illustrated in Susan Griffin's distinction between two senses of grasping. One is "to *seize,* and *grip,* as in *wrest power from the grasp of* or *grasp a woman by her waist*" (Griffin 1992: 212, emphasis in original). This is the power of dominion, the commanding grip or the judging gaze. The other way of understanding is enacted more by a mobile glance than a fixating gaze (Jay 1994: 56–57). It lies in grasping a truth that is "a delicate gesture, like taking a hand in greeting. A lightness of touch is needed if one is to feel the presence of another being" (Griffin 1992: 212). The lightness of the touch (of the Other) was precisely the boundary-crossing simultaneously most desired and feared by the missionary males and their grasping fellow scribes who sought the seductive promise of encounters with difference while simultaneously pushing the frightening difference to the forbidden category of absolute Other (see Connolly 1991). By way of shoring up their nearly breached perimeters, they overzealously imprinted their truth, an "excess of gender hierarchy," on the realm of dark Hawai'i.

This kind of discursive violence enacted in the colonizers' complacent acceptance and rationalization of genocide continues today in the common argument that, since the diseases were not introduced "on purpose," and since it is plain that we "cannot turn the clock back," the best that we can say is that it was a regrettable accident and we must "put it behind us." To make historical roadkill of the Hawaiians is to deny as well as perpetuate the violence of the discoverers' feminization of the land and its recapitulation in The Mission's heathenization of its people.

PLANTERS

There were no dockside blessings, no instructions issued to the men whose uncoordinated, unpredetermined arrivals ultimately produced a Hawai'i extravagantly respatialized for monocrop agribusiness and a social formation that put racialized bodies productively to work. Race became an active, marked, and socially subordinate category, not one potentially to be refigured white, as the missionary narrative had it. As men the planters *were* their own narrative, a mid-nineteenth-century version of "We Are the World," embodying some of McClintock's "stalwart themes of colonial discourse: the feminizing of the land, the myth of the empty lands, the crisis of origins . . . the ordering of land and labor, the invention of racial idleness" (McClintock 1995: 17). Most were Americans; some were educated; a few were missionaries and missionary sons; others were seedy adventurers, small-time entrepreneurs, men with a self-reflecting eye on the main chance, all drawn to the promise of "virgin lands." Illustrating yet another manifestation of the tension between anxiety and desire in the colonial gaze, the agricultural entrepreneurs both claimed that the land was already "empty" and participated in the efforts to empty it. The desire to both secure and normalize this emptiness, employing the twin alibis of natural order and divine intent, required them to work energetically on the physical and human landscape to produce an order that they considered to be natural and universally valid in the first place. The myth of the empty virgin lands made them an offer they could not refuse:

> Within patriarchal narratives, to be virgin is to be empty of desire and void of sexual agency, passively awaiting the thrusting, male insemination of history, language and reason. Within colonial narratives, the eroticizing of "virgin" space also effects a territorial appropriation, for if the land is virgin, colonized peoples cannot claim aboriginal territorial rights and white male patrimony is violently assured as the sexual and military insemination of an interior void. (McClintock 1995: 30)

Demonstrably not quite emptied, the land was not quite freely available either. Chiefs might permit foreigners to use the land but their notion of permission was not confined by the guarantees of boundedness, access, and tenure that went with "proper" title to it. The relationship of Hawaiians to land was not an alienable one; the divisions and uses of the land, to the western eye, were apparently disorderly.[24] Divided into *ahupua'a* (land wedges running from ocean to mountain top), land boundaries wandered from a rock outcropping to a clump of bushes to a proud tree and other natural landmarks and on out to the reef. As the land emptied of Hawaiian bodies, some drawn to Honolulu and the several other cities created by westerners, some drafted to collect sandalwood from the mountains by their grasping chiefs, and more to their deaths, the cultivation of *kalo,* a basic element of the nourishing fish and poi native diet, declined, and the *lo'i* became overgrown with grass.[25] Persuaded by western "advisers" and some small-scale sugar growers, both of whose numbers included missionaries, that they must realign human relationships to the land, Hawaiian rulers enacted the *Māhele,* the dividing of the land, at midcentury. Once land was commodified, its borders violently straightened and straitened by surveys, titles, and land court rulings, another stream of order began to be settled on the land. The casual border crossing of survey lines became illegal.

By August 1850, the motley of men had become a collective, the Royal Hawaiian Agricultural Society. Through its meticulous organization of committees—including committees on labor, roads and harbors, soil analysis, coffee and cane, trees and grass, poultry, sheep, seasons, insectivorous birds, cutworm, and other "destructive insects and vermin"—and through its full-court press on the government—to fund agriculture, make a geological survey of the islands, facilitate the importation of agricultural machinery, establish a public nursery, and in general begin a program of funding an economic infrastructure—the Royal Hawaiian Agricultural Society initiated the possibilities and rules of articulation for a different discourse on the social formation that was Hawai'i (*Transactions* 1850: 17). The society created itself from men responding to invitations sent to "all Farmers, Planters, Gardeners, and other persons interested in the formation of a society for the promotion of Hawaiian Agriculture" (*Transactions* 1850: 4). An interminable inaugural address by Judge William L. Lee laid out the capitalist economic order, beginning with a telic narrative of agricultural possibilities in Hawai'i enabled by the feminization of land and its now empty (as of the *Māhele*) state:

> Perhaps there is no country in the world that offers greater promise to the husbandman and grazier, at the present time, than the Sandwich Islands. The lands are now thrown open to all classes, the native and foreigner, the subject and alien. We have a climate unparalleled in its salubrity, and affording every variety from the perpetual snows of Mauna Kea to the burning plains of Waikiki. Our soil, though not deep, is warm, quick, and fertile. (*Transactions* 1850: 3)

Staking out agriculture as standing "first in importance and second to none in dignity" (*Transactions* 1850: 25), Lee's survey of agricultural civilizations of the past, from Egypt to Greece to Rome . . . to Hawai'i, was the instillation of a "geography of power" (McClintock 1995: 37). Barely able to restrain his enjoyment of anticipated delights and riches to issue from the land, Lee interrupted his beat to remember that "the soil whose treasures we would unlock was once the undisputed heritage of the poor Hawaiian" and that it was white men whose diseases are driving Hawaiians into extinction (*Transactions* 1850: 35). Recovering his rhythm within several sentences, he employed the familiar protoroadkill justification, arguing that if the efforts of Agriculture do not send a "quickening pulse through the heart of a wasting nation" (and save the remnants of its people), then "we can but commend it to Him in whose hands are the issues of life and death,—to Him who 'counteth the nations as the small dust of the balance, and who taketh up the Isles as a very little thing'" (*Transactions* 1850: 35). Having threaded his way past the sticky question of origins and implicitly blamed Hawaiians for their susceptibility to western diseases, Lee finally gathered himself for the question of just who was, in fact, going to do the work of agriculture: "there is one agent, however, that we require, who holds the key to success,—the great brawny-armed, huge-fisted giant called LABOR" (*Transactions* 1850: 33). Coded masculine, singular, "Labor" functioned as a necessary but unpredictable monster, capable of awesome power while requiring continuous control:

> We must not only secure him, but direct, aid, and economise his efforts; for though with his strong arm and heavy hand he is might to accomplish, yet unguided by the art of man he strikes like the blinded Cyclops of old, destroying with one blow what he effects with another. (*Transactions* 1850: 33–34)

With these words, Lee began to articulate what was to be a color-coded hierarchic economic order. It soon become clear that the actual labor of agriculture, even if "second to none in dignity," was dead last in desirability in the white man's eyes. Fortunately, others were available to take the role of Cyclops.

First to be brought as contract laborers were the "coolies from China" in 1851 (*Transactions* 1851: 91). Two years later, the manager of Līhuʻe Plantation reported that

> we continue to like our Coolies. The number is now increased to 41. . . . As laborers in the field, the Coolies give us perfect satisfaction. . . . they are industrious, skillful, and thorough, and one Coolie in a cane field is worth in my opinion three natives. But among themselves, they are quarrelsome and passionate and have many dangerous fights. (*Transactions* 1853: 123–24)

Subsequent labor committee reports continued to elaborate the theme of the idle natives ("the aboriginal population could not be depended on as laborers") and urged the acquisition of more foreign laborers, including the importation of "women as well as men from China" (*Transactions* 1853: 140–41).[26] The anticipated benefits of supplementing the labor hierarchy with women were hardly surprising: a supply of domestic servants, labor for "light" duties around the plantation, and, through marriage, "a new element of population which may re-invigorate the native race, or if the latter be so effete as to be incapable of restoration, supersede it" (*Transactions* 1853: 140–41). Either way, through Hawaiians intermarrying with the Chinese women and reproducing a racially mixed labor force, or the "effete" Hawaiians hitting the end of the reproductive trail and leaving the reproduction of laboring bodies to the Chinese, the planters thought themselves covered. In their narrative, they would direct Cyclops and Cyclops would enrich them.[27]

Guaranteeing themselves a sufficient supply of laborers was a perennial "problem" for the planters that caused two related anxieties for them. One was the agency of the workers themselves; some returned to their homelands after serving out their contracts, and some refused field work. "Ambition" spurred others such as the Portuguese to search for other occupations and "ris[e] on the ladder of success" (Hawaiian Sugar Planters' Association 1921: 18). Japanese laborers reached across ethnic lines to Filipinos in 1920 to strike for higher wages.[28] Such anxieties were exacerbated after Annexation by several factors: American immigration laws (the Chinese Exclusion Law and the Gentlemens' Agreement) that cut off the supply of Asian labor; efforts by the Philippine government to restrict emigration of its men to Hawai'i; and criticism, both local and mainland, of the Masters and Servants Act and of labor conditions on the plantation.[29] The other side of the anxiety about an ample supply of color-coded laborers was the fear of engulfment. "The White Man in the Tropics," and his "natural" unsuitability for "hard, heavy outdoor work under a tropical sun," a theme repeatedly announced to themselves, exhibited their sense of boundary insecurity (Hawaiian Sugar Planters' Association 1921: 31). Their attempts to neutralize it were unceasing and took many forms. Laborers from different countries were kept apart on the plantations; their overseers were selected from different ethnic groups. Planters' correspondence and publications obsessively dwelled on the relative merits and, principally, demerits of the different "races."[30] These themes were later to be given some standing by a branch of social psychologists including Stanley D. Porteus, who, in a work largely addressed to the Caucasian audience in Hawai'i, wrote that it was possible to measure the "intelligence" and "temperament" of diverse groups of persons—"races"—whom he rather loosely termed as Chinese, Filipino, part-Hawaiian, Japanese, American, and so forth. His measurable characteristics included such terms as *impulsiveness, deliberativeness, emotionality, suggestibility, volatility, heedlessness, an ability to plan,* and *confidence.* Porteus found that some races naturally had lower plateaus of ability than others, but that the Japanese had inherent abilities that challenged those of Americans. Consequently, he recommended excluding Japanese from entry to the United States (Porteus and Babcock 1926).[31]

Having deployed race as a means of securing their economic domination, the planters also faced the return of the racially repressed. Since the largest segment of the 1921 population of Hawai'i was Japanese (41 percent), while the white population consisting of Americans, British, Germans, and Russians combined was only 13.5 percent, the numerical engulfment of white bodies by colored ones had already occurred (Hawaiian Sugar Planters' Association 1921: 19). The predominance of Japanese in the labor force constituted an equally serious assault on the planters' containment reverie; since nearly half of the total of 39,500 field workers were Japanese, it was "easy [for the planters] to see how the Japanese, welded together in racial solidarity, possess a potential power of economic domination" (Hawaiian Sugar Planters' Association 1921: 21).

There is nothing like colonizing power, however, for the resources and means that it affords for endlessly inventing new strategies of containment and control. Able to attract Filipinos to the fields, to control the territorial government whose governor was federally appointed, and to mobilize a dense network of interlocking directorates of banks, plantations, and infrastructural institutions, the planters embarked on a refiguration of the alien bodies—their co-optation and Americanization. The latter was and remains a visible, explicit project of the public schools.[32] A striking semiotic text of the co-optation project of the planters is a narrative related by the Hawaiian Sugar Planters' Association in a series of twenty-six weekly advertising vignettes appearing in a local newspaper in 1934. In these "public statements," the sugar industry sought to present "Hawaii's story to Hawaii's people" as an accomplishment "brought about through a spirit of cooperation." It was an unselfconscious narrative eliding any boundaries between "Hawaii" and the sugar industry, reinstalling a telic order of gender, race, and class while unreeling "the story that needs telling." The planters concluded that "what has happened . . . and caused the world to pause and ponder . . . could never have happened in any other way" (*Honolulu Star-Bulletin* 1934: 7). Race and class know their place in the illustrations; the white man still does not labor in the tropics. The logo appearing on each page is two figures in straw hats, long sleeves, gloves, and pants leaning in to their job of cutting cane with cane knives (fig. 3). Other male laborers carry sacks, talk over the back fences of their plantation cottages, work as stevedores at the docks.

Figure 3. Drawing of workers hand-cutting cane, used as a logo by the Hawaiian Sugar Planters' Association in 1934 in a newspaper advertising series.

The women, who also worked as field hands, served as domestic servants, and, as always, made men's work possible and reproduced the labor force, are invisible, but different women, white of course, are mothers, sales clerks, and teachers. The Men with Heavy Responsibilities wear suits and grave looks; other white men run newspapers and work in laboratories. There is a suggestion of a multiethnic middle class (of seven figures, one woman is dressed in a kimono and a couple in the foreground could be Asian as well) in front of a building, appropriately labeled "Bank." Still other illustrations feature the power of industrial technology: cranes, harbors, road-building equipment, iron foundry. And then there is fertile Nature—sunshine, moonlight, a stalk of sugarcane, waterfalls, and a rainbow. Tropical tropoi come thick and fast here; Hawai'i's sunshine and climate are golden, the green stalk reaching for the sun is sugar, and the rainbow's promise is fulfilled: "The wealth [sugar] has attracted to Hawaii has made possible a community life as fine and modern as any in the world. . . . A stalk of cane . . . the sugar industry . . . Hawaii's crock of gold" (*Honolulu Star-Bulletin* 1934: 7).

The standardization of labor sought by the planters was part of a "conversion project, dedicated to transforming the earth into a single economic currency, a single pedigree of history and a universal standard of cultural value" (McClintock 1995: 34). This "global science of the surface" (McClintock 1995: 34) recruited dark bodies to be injected as human equipment into the "warm, quick, and fertile" soil that Judge Lee so fervently praised. The landscape as the planters taught it to speak required a class of laborers who simultaneously confirmed and threatened sugar's hegemony. The complex comings and goings of discovering, settling, seizing, and laboring bodies help to explain the complexities and sensitivities surrounding contemporary constructions of identity and place. Benign metaphors of "melting pot" or "rainbow" depend on collective amnesia to deny the political economies of immigration, the color-coded categories that marked some incoming bodies as entitled to conquer or convert, and others as destined to serve, or perhaps just to get out of the way. Hawai'i's much praised multicultural present has its inception in international trade in bodies marked by their color for subordination.

MILITARY TRAFFIC

Waves wash up on beaches and then recede; some of the flotsam is carried up far enough to resist the suction of the water taking the rest back into the ocean. Things deposited on the shore are subject to continual suction, but some also become embedded in the sand. There is rarely a specific moment in the usual work of the ocean when the permanent embedding of an object can be determined; rather, that occurs over time. Similarly, it is difficult to name a moment when the military order became embedded in Hawai'i. It occurred in a series of developments, "moments of intensity . . . lapses . . . periods of feverish agitation . . . fainting spells" (Foucault 1977: 145). Ships, not flotsam, carried the first intimations of such an order. There were British Marines aboard Cook's naval ships in 1778; both Cook and George Vancouver were commissioned officers in the Royal Navy, and Wilkes in the U.S. Navy. Both British officers gave western military technology to Kamehameha I; British ships visting a few years later carried on a regular arms trade with various chiefs. Trading ships usually were armed as well. One notable skirmish involved an American trader who, when a small boat attached to his ship was stolen during the night, lured the natives from the shore the next day and, when they drew close, fired rounds of shells into them, killing more than a hundred, wounding many more. A few weeks later, his son, apparently knowing nothing of the event, sailed a small schooner into waters off Kona and was gulled into permitting Hawaiians on board his ship. They seized the vessel and killed all the crew save one. This survivor, Isaac Davis, along with an English sailor, John Young, ultimately became part of Kamehameha's entourage; autodidacts though they were, these bits of flotsam imparted other martial and technological knowledge to the Hawaiians (Kuykendall 1957: 24–27).

By 1810, the protected harbor of Honolulu was becoming commercially important: firearms, household utensils, cloth, pigs, fruit trees, alcohol, iron tools, fleas, mosquitoes, centipedes flowed in. Many Hawaiian bodies flowed out to become hands and occasionally masters on commercial and whaling ships. Sailors and whalers flowed in; riots occurred occasionally as the men demonstrated violently against the attempts by missionaries to prohibit prostitution in the ports of Honolulu and Lāhainā (Kuykendall 1957: 122–23). U.S. Navy ships were sent on several occasions as bill collectors for American merchants, creditors of various chiefs who had pledged payment in sandalwood and, later, land. After 1826, American warships began regularly to visit Hawai'i (Kuykendall 1957: 91–92). They were not alone, as British and French warships also made their presence known but less regularly.[33] In the decade following the American Civil War, American warships, now numerous enough to constitute a Pacific squadron, called more often. That they should, Kuykendall reports, was not a new idea, but a "common thought in the minds of Americans resident in the islands and of many in the United States" (Kuykendall 1966: 207). The *Lackawanna,* commanded by Captain William Reynolds and reporting to North Pacific Squadron commander Rear Admiral H. K. Thatcher in San Francisco, was assigned indefinitely to Hawaiian waters, in order "to guard the interests of your government faithfully and give all proper protection to its citizens abroad" (Secretary of Navy Welles, quoted in Kuykendall 1966: 208).

At the same time, pressure began to increase for American annexation of Hawai'i and for a reciprocity treaty between Hawai'i and America. A new American minister, Edward McCook, was appointed to Hawai'i, an early step toward what has become a common American practice of the "revolving door"—the circulation of military bodies through both public and private governing boards. McCook rather soon became a proponent of the move to secure a reciprocity treaty with the United States, a treaty much desired by many of the sugar planters, whose exports to the large American market were subject to a tariff. Opposition came from other planters who wanted *all,* rather than some, grades of sugar admitted duty free, from those native Hawaiians who opposed increasing American influence in Hawai'i, and from those who wanted Hawai'i annexed by the United States. A further obstacle was the continued presence in the Honolulu Harbor of the *Lackawanna.* Kuykendall explains that although the ship had been assigned to Hawaiian waters to forestall aggressive French action against the Hawaiian government, it was seen by the Hawaiian government as foreign surveillance of themselves (Kuykendall 1966: 209–16). Spying by foreign governments, while not completely the order of the day, took place regularly enough to make it clear that the early superpowers were learning their trade and that in their eyes Hawai'i was an inferior power. Another American military man, recycled as a supposed cotton planter, was actually a courier for Secretary of State William Henry Seward. In 1868, a short time later, he became a vice consul and then acting chargé d'affaires of the U.S. delegation. When the Hawaiian government came to know of his secret activities, they refused to acknowledge him as a consular officer, but, when he was given a regular consular commission later in the year, they were forced to accept him.

Other American military intrigues of that time included the use of 150 U.S. Marines in 1874 to quell protests that occurred after an election in which Kalākaua was elected king, defeating Queen Emma whose supporters were causing the affray (Daws 1968: 198–99). Probably the most well-known spy of the time was General John M.

Schofield, who along with General B. S. Alexander arrived in Honolulu in 1873 on the *California,* the flagship of Admiral A. M. Pennock,
now commanding the U.S. North Pacific Squadron. The *California*
had been ordered to Honolulu to escort King Kamehameha V to San
Francisco. Although he died before the ship left San Francisco, the
navy did not countermand its sailing orders and directed Admiral
Pennock to "use all your influence and proper means" to influence
the king's successor to look with favor on American interests (Secretary of Navy Robeson, quoted in Kuykendall 1966: 248). Schofield
and Alexander, however, were not accidental travelers. Purportedly
on a vacation to allow Schofield to recover from poor health, the
two were under secret orders from U.S. Secretary of War Belknap to

> ascertain . . . the defensive capabilities of the different ports and their
> commercial facilities, and to examine into any other subjects that
> may occur to you as desirable, in order to collect all information that
> would be of service to the country in the event of war with a powerful
> maritime nation. . . . It is believed the objects of this visit to the Sand-
> wich Islands will be best accomplished, if your visit is regarded as a
> pleasure excursion, which may be joined in by your citizen friends.
> (Belknap, quoted in Kuykendall 1966: 248)

The two well-concealed spies remained for two months, receiving
much assistance from local authorities. Their report, made public
twenty years later, emphasized the value of Pearl Harbor and dis-
cussed means of enlarging it for naval and commercial purposes. The
surgery proposed by Wilkes to provide for easier penetration by U.S.
naval vessels was moving closer to realization.[34]

The military gaze on Hawai'i recruited the informational and relational gazes entirely in the service of the gaze that appropriates and
controls. What could be said about Hawai'i, and what could be seen
in Hawai'i, collapsed around and enabled the constitution of what
could be seized. What could be seen in Pearl Harbor, for example,
did not include the extensive system of fishponds central to the political economy of native life. The desire in the military's eyes seems a
bit calmer, less frantic, more matter-of-fact than that of the missionaries, less greedy than that of the planters. But the facts that matter
are those that confirm Hawai'i's availability and America's entitlement, Hawai'i's inadequacy and America's need to take charge.

The Reciprocity Treaty finally signed in 1876, a centennial year for the United States, allowed all grades of Hawaiian sugar into the American market duty free, a tremendous relief for the planters who had described themselves as "facing disaster." The Hawaiian kingdom had been able to resist ceding Pearl Harbor to the United States but was bound to promise not to allow any other foreign powers to acquire property in Hawai'i (Daws 1968: 202–3). By the time the treaty was renewed in 1887, planters and other businessmen, largely American, had forced a new constitution on the king (the Bayonet Constitution). It enabled them to move into controlling positions in the Hawaiian government and left the king with few means of effectively opposing the new treaty, which granted the United States exclusive access to the resources of Pearl Harbor (Lind 1984/85: 28). The U.S. Congress had insisted on this prerogative as a price for negotiating another treaty (Kuykendall 1967: 373–400). With Pearl Harbor secured, only one more step needed to be taken before landing of military bodies firmly and finally on the shore could begin—the annexation of Hawai'i.

At his death in 1891, King Kalākaua was succeeded by his sister, Lili'uokalani, who made known her dissatisfaction with foreigners' control of the government of Hawai'i and announced plans to promulgate a new constitution restoring the power of the monarch and cabinet. This precipitated a panic among the thirty or so lawyers, editors, planters, bankers, businessmen, and other men who had in so many ways pushed and pulled the fabric of Hawai'i into a shape that fit them comfortably. Counting on the prior, if secret, promise by the American minister to support with American troops their efforts to dethrone the queen, the dozen or so men who made up the Committee of Safety[35] asked the minister for armed intervention because public safety, lives, and property had been endangered by Lili'uokalani's announced intention. As promised, four boatloads of marines from the *Boston* were landed; the positions they took up were considerably closer to government buildings than to the lives and property of those claimed to be in danger (fig. 4). To avoid armed combat, the queen surrendered her power in the belief that the U.S. government, upon learning the facts, would restore her and her government to power.[36]

American "destiny" for Hawai'i was played out in 1898, even as it was being played out in the declaration of war by the United States with Spain over Cuba and the Philippines. Naval captain and strate-

Figure 4. Marines from the USS Boston *in support of the overthrow of Queen Liliʻuokalani. Courtesy Hawaiʻi State Archives.*

gist Alfred T. Mahan had already declared that American destiny was linked to the "fundamental truth . . . that the control of the seas, and especially along the great lines drawn by national interest or national commerce, is the chief among the material elements in the power and prosperity of nations" (Mahan 1918: 285). Mahan firmly believed in the annexation of Hawaiʻi owing to the predominant interests of the United States in the region. Other military voices calling for American annexation included the well-known spy of an earlier decade, General John M. Schofield, as well as Admiral Lester A. Beardslee, commanding the U.S. Squadron in the Pacific, and Colonel Charles Eagan, Assistant Commissary General of Subsistence, U.S. Army. Eagan in particular argued eloquently that there should be opportunities for white men in Hawaiʻi, that other nations wanted Hawaiʻi ("No one can justly blame them for this desire"), that the Hawaiian race was dying out, and that it was the duty of the United States to say "Hands off Hawaiʻi" to all other nations. Pearl Harbor, he felt, would make a handy coaling station for the modern

American man-of-war, without which "she is merely so much iron and steel, harmless as the proverbial lamb" ("Important Views" 1897: 8). No one asked if there might be a different concept of duty that would require the United States to restore Hawai'i's sovereignty and to assist the islands in protecting their independence. The claims to national sovereignty and autonomy that the United States and Europe took for granted for themselves were not part of the story the superpowers told about Hawai'i, because Hawai'i, with other colonized places, was taken to exist in a different time, a suspended time, a time outside of proper History (McClintock 1995: 40). American sovereignty confirmed the march of "progress . . . the footstep of God himself" (Victor Hugo, quoted in McClintock 1995: 10), while in Hawai'i's case it was the destruction of sovereignty that confirmed natural order and divine intent.

Military inscription on Hawai'i's spaces was deep, extensive, and immediate following Annexation. The military buildup began four days after Annexation with an initial company of soldiers garrisoned near Diamond Head. The construction of a naval base at Pearl Harbor began with dredging in 1900; the first warship entered in 1905. Formal opening ceremonies took place on December 14, 1911 (Ramsey 1982: 3). Forts Shafter, Armstrong, DeRussy, Ruger, Kamehameha, and Weaver and Schofield Barracks were built before the First World War began (fig. 5). The Manly Military Body had arrived to defend its Hawai'i by means of "a ring of steel, with mortar batteries at Diamond Head, big guns at Waikiki and Pearl Harbor, and a series of redoubts from Koko Head around the island to Waianae" (General Macomb, quoted in Lind 1984/85: 25). The early semioticians of militarization, like those of discovery, Christianity, and the plantation economy, translated Hawai'i into a cultural and physical space they then read as inviting the order they imposed on it.

Over the years, military order has become naturalized, rendered ordinary and unremarkable in establishment discourse. The widespread acceptance of the military as an industry by both business and government leaders, and as just another job by many people, suggests that the work of the early semioticians has had a lasting effect. A Bank of Hawaii research publication reviewing the impact of the military on the state economy found that "during the past 35 years the military registered the most consistent growth pattern of

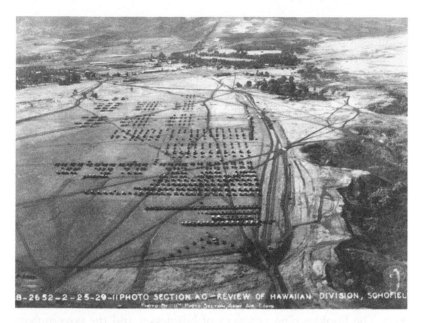

Figure 5. The carving out of Schofield Barracks on the Leilehua plain, central O'ahu. Courtesy Hawai'i State Archives.

the major segments of the islands' economy" (1984: 1). It calculates that military spending in the local economy has a multiplier effect of three, that is, that every dollar spent by the military turns around three times in the local economy. The bank's report concludes that, despite the difficulty in measuring economic and social costs, the "military population tends to offset the costs of its presence in a way that others cannot" (3). A more recent study by First Hawaiian Bank refers to the military as "one of the brighter aspects of our economy" because Hawai'i has suffered fewer cuts by the Base Realignment and Closure Commission than any other state (1995: 3). Explicitly invoking both threat and bribe, the study invites citizens of Hawai'i to compare our relatively prosperous situation to the economic hardships imposed on Guam by military cuts, using Guam as an example "of what can happen to a defense-dependent island economy in this environment" (3). Interpreting the military's history in Hawai'i as an innocent economic contribution rather than a colonizing presence, the bank urges the state to "continue to build on its heritage as an island whose economy was in tangible part built by military injections earlier in this century" (First Hawaiian Bank 1995: 3). The National Commission for Economic Conversion and Disarmament concurs with the two banks' findings, showing Hawai'i to be one of the few states to receive more per capita from military spending than it pays in its share of taxes to the Pentagon (1995: 4).

Such studies carry great weight in Hawai'i for obvious political and economic reasons. However, it is not clear from these studies how much of the military's appropriated spending actually goes to local payrolls and suppliers. Reckoning the relative costs and benefits of the military as an industry is an admittedly complex operation. There are reasons to regard the question as at least an open one, given the paucity of second opinions. The state is dependent on military sources for data on military numbers so its studies always carry this (often unacknowledged) debt (interview with Pfister 1991). Moreover, military advisory committees are drawn from the ranks of leaders from corporate, banking, civic, educational, political, and social groups; this hegemonic group is not likely to question its own base. The bookkeeping practices of businesses and the government have drawn tight boundaries around the term *cost,* limiting it to certain kinds of financial data and precluding such social, economic, and environmental costs as native species destroyed, soil and drinking water polluted, land destroyed by military operations, housing costs driven up by rental subsidies for military personnel, *heiau* (temples) destroyed by federal highway construction, fishing grounds dredged for harbor expansion—the countless insults to local ways of life by self-proclaimed military entitlements.

Barrages of claims about military economic contributions come at Hawai'i residents on nearly a daily basis, with relatively little information or argument going in the other direction (see chapter 6). Even to suggest the possibility of such arguments is often taken, in elite circles, as somewhere between seditious and absurd. University of Hawai'i student M. L. Kilgore details his account with the sacred:

> At the State Democratic Convention, I helped to floormanage the passage of an amendment to the platform dealing with the need to diversify the economy to achieve greater self-sufficiency.[37] The original wording of the article, which was rejected by the platform committee, contained the phrase "rechannel resources toward peaceful enterprises and away from the military." When this wording was deleted, the Chair of the platform committee rose to speak in favor of adopt-

ing the amendment with "the extremely objectionable wording now removed." The incident drove home to me the lesson that in Hawaii the military is sacrosanct. (Kilgore n.d.: 1)

In a rare public discussion of the possibilities of converting Pearl Harbor to nonmilitary production, the governor's special assistant to economic affairs demurred, fearing to give offense to military officials: "You don't want to start creating alarm by going out and saying, 'Oh, looks like the military is going to decline, and so we're going to make these alternative politics.' In a sense, that's like a slap in the face and a vote of no confidence in the military" (Seto 1990a: A-4). In a similarly rare public acknowledgment of the closed discourse on the military's costs in Hawai'i, State Representative David Hagino decried "a military-industrial mentality that refuses to let go. There are two industries (tourism and military) that are so big that they cloud our judgement and perspectives on the future" (Seto 1990b: A-8). In light of this clouded judgment and of the many social and environmental costs, not usually included in the calculations, the economic cost-benefit ratio is an open question. More broadly, there is an ontological violence in the ruthless erasure of the ways of life and land incompatible with military mandates. "National security" is the military's trump card, used successfully most of the time, to move people, claim resources, take land, install and extend always new and better technologies of war. The political resources needed to rethink this claim to security, to challenge this kind of order, surface most visibly in Native Hawaiian communities and among environmental, feminist, and peace activists but are not widely available in the larger population. The paucity of such political resources indicates the costs imposed by a militarized society on active, contestatory, democratic practices of citizenship.

THE TOURISTS

All encounters with Hawai'i so far described have been led by the colonizing gaze, that restless mixture of anxiety and desire, from the first voyagers to the missionaries, the planters, the professional soldiers, and to the millions of persons who now arrive each year. For some, it was a spatialized gaze that sought something, usually an empty, feminized, always potentially pregnant land. For others, it was a gaze that sought someone, a peopled land: people who were interestingly different but vulnerable to the outside world; people hopelessly and disgustingly different; people indifferent to work and, even more, to their own survival; people incapable of governing or defending themselves. Not all the functions of the gaze—optical, linguistic, haptic—are equally present in the preceding encounters with Hawai'i. But the tourist gaze is inclusive; the gaze informs, gazes are exchanged, and by the gaze I touch, I attain, I seize, I am seized. "The gaze is an anxious sign: singular dynamics for a sign: its power overflows it" (Barthes 1991: 238).

Barring Cook and his men, there was not an identifiable first tourist. But, to cut to the chase, tourism became possible after 1867 when the United States Postmaster General contracted with the California, Oregon, and Mexico Steamship Company for the monthly steamship delivery of mail to Honolulu from San Francisco. Thereafter, regular steamship service continued nearly uninterrupted until the Second World War (Kuykendall 1966: 170), after which it was superseded by the airplane. The government was prompted by the establishment of transportation and communication with San Francisco and with the British colonies in the south to construct a hotel to provide for a "large increase of visitors" as well as for those who might be anticipated in the future (Kuykendall 1966: 172).[38] By 1903, the Chamber of Commerce and Merchants Association created a Hawaii Promotion Committee, a first step in a new spatialization of the land of Hawai'i (Goss 1993). The refiguration of laboring and visiting bodies was necessary to interpellate a new kind of border crosser, to hail new subjects onto the ideological horizon (Althusser 1971: 174). Photos of Waikīkī beach in the early twentieth century show that Hawaiians and other local residents still had access to the ocean and littoral. Under the impetus of the new accommodating (to tourists) order, the Hawaiians fishing with throw nets and outrigger canoes disappeared as indigenous users of the beach and ocean. They came to be replaced by the domesticated gentle male Hawaiians of "splendid physiques" and "dignity of bearing" who began to make the ocean safe for tourists (Ferdinand J. Schnack, quoted in Herman 1995: 271). These new Hawaiians paddled canoes and taught tourists to surf; they strung leis, sang, and strummed ukuleles.[39] Some Hawaiian men faded into quaintness (The Hawaiian Boys Diving for Coins Tourists Threw into the Harbor from the Deck of the Ship; The Fisherman Casting His Net; The

Native Climbing the Coconut Tree Barefoot). Other local men moved from agricultural work on the plantation to agricultural work on hotel grounds or to other out-of-sight work in the kitchens and similarly invisible spaces in tourism's social production.

The tourist industry sees Hawai'i as a place that promises exotic leisure in a setting that is nevertheless safe and marked with familiar cultural symbols. The opportunity to enjoy blue skies, balmy weather, endless beaches, and intriguing local color is realized by "development"—the economic transformations that produce high-rise hotels, time-shared condominiums, chain restaurants, international car rental companies, golf courses, standardized entertainment featuring exotic local performers—the accoutrements of McLeisure. Where dark-skinned workers were formerly recruited for plantation labor, they are now hailed into place as the service providers in the international political economies of tourism.

Hawaiian women at the turn of the century also began to be reconfigured; they were photographed bare-breasted, wearing grass skirts, possibly leis. Early photographers never tired of photographing Hawaiians, occasionally Asians, both in "native costumes" and western dress: "Hawaiian House, Hana" (typical thatched house with three figures seated nearby); "Lei Sellers, Honolulu" (five females wearing "colorful" full-length garments); "Hawaiian Family at a Traditional House, 1890" (a thatched house with a number of adult figures standing and sitting in front of it holding young children); "Hawaiian Woman, Hana" (a studio photo of a Hawaiian woman dressed in a full-length garment and holding a fan downward); "Hawaiian Couple, Hana" (another studio photo with the woman seated wearing a full-length garment while her husband, dressed in western-style suit, stands by her side with his hand just touching her shoulder); "Distinguished Hawaiian Family, Hana" (a middle-aged Hawaiian couple, she dressed in enveloping western clothes, he in a suit, seated on chairs outdoors, with a small boy, dressed in western finery, perched between them); "Hawaiian Couple in Grass Home" (a rarer interior view of a thatched house; the couple is dressed in western-style clothing, sitting on mats behind several "cultural objects" such as a calabash, perhaps a lauhala basket arranged in front of them) (Davis and Foster 1988).[40] In none of these photos is the gaze reciprocal; rather, the desiring gaze of the camera records the social objectivity of its defeated subjects.

The trajectory of Hawaiian women as objects of desire has run a gamut of redefinition. Taken as generally unattractive by discoverers (even as the men demanded access to the women), they were rewritten as threateningly wild, lewd, and licentious by the missionaries and later dismissed as dulled creatures by prospective employers, who saw the women as difficult to train to household tasks (Herman 1995: 16–18).[41] Yet later, as R. Douglas Herman remarks, "with the ascension of white power in the Islands, Hawaiian women—like the Hawaiian men 'of old'—bec[a]me increasingly beautiful," often, particularly, the hapa-haole (half-white) women (1995: 179). The ascension of white male power had not only spelled out safety for the sugar industry and the military, but it began to underwrite the sexual fantasies and social practices of tourism as well. Desire and anxiety worked together to create the exotic/erotic Other. Feminized Hawaiian males, desexed Asian menials, and exoticized Hawaiian "hula-hula" girls constituted the sturdy labor base and the refigured subjects for this new order.[42] As giant concrete "jacks" are locked together and set in place to stabilize a beach or prevent reef erosion, race, class, and gender were interdigitated in a new configuration to provide profits for the tourist industry.

This new order revalued certain kinds of land. While it still needed to be "empty," it was no longer measured by its productivity in metric tons or contiguous acres, but by its proximity to sandy beaches and clean bays. The physical borders of the islands became the new prizes while the corollaries of blue sky, warm weather, and brilliant sun did not need to be hailed into existence for the new traffic in bodies to begin. The coastline and beaches could always be emptied through purchase, if necessary, or through interpretation of laws, or fine-tuning of the community-planning agencies. And if the uncooperative ocean, following its own logic of deposits and withdrawals of sand, redistributed some of the resources, it was always possible, with the right engineering technology, to ferry it back again. If a reef or a rock offended tourists' toes or eyes, it could be plucked out. A dredge here, a dredge there, and a lagoon appeared where none had been before. "Tropical" foliage came into existence, often lowered into place by crane or planted and carefully maintained by silent workmen, silenced by their Asian origins and their position in the political economy of tourism. All were laboring to produce the movie *Romance in Paradise with the Full Moon Shining on the Water and the Waves Lapping Softly at the Shore,* starring the Sinuously Exotic Native Dancer and the Friendly Hawaiian Male with a Smile on His Face and a Song in His Heart.

Jon Goss has identified the tropes paradise, marginality, liminality, femininity, and aloha as the "persistent themes of the spatializing discourse" found in the "destination marketing" of Hawai'i (Goss 1993: 663). While Goss has restrained his study to two recent decades, the tropes expressed by "green, verdant, lush, rich, and fertile" did not suddenly emerge from hiding in 1972 but, rather, are derivatives of the haptic gaze focused on Hawai'i from the time of contact (Goss 1993: 676). They are the tropes that sell.[43] They are also the tropes that shape the land as well as excite the desires. "The metaphor," Hayden White remarks,

> does not image the thing it seeks to characterize, *it gives directions* for finding the set of images that are intended to be associated with that thing. It functions as a symbol, rather than as a sign: which is to say that it does not give us either a *description* or an *icon* of the thing it represents, but *tells us* what images to look for in our culturally encoded experience in order to determine how we *should feel* about the thing represented. (White 1978: 91, emphasis in original)

Feminine metaphors of warmth, fertility, and ripeness; racial metaphors of darkness and difference; class metaphors of leisure and plenty—these operate within discourse as heuristic devices producing the desired gender-effect, race-effect, and property-effect. They prefigure the tourists' field of perception and interaction so that only some information is sought (not, for example, the appalling statistics on the high percentages of Hawaiian men in prison), only certain relations are cultivated (not, for instance, with those in the sovereignty movement who detest Hawai'i's dependence on tourist dollars); and, above all, so that no challenge is raised to the visitors' entitlement to consume pleasures and products that have not always been viewed as commodities.

The "mobile army of metaphors, metonymies, anthropomorphisms" that has played across the history of Hawai'i's militarized present configures and reconfigures the tropological territory through the engines of anxiety and desire (Nietzsche 1956b: 180). The gazes of discoverers, missionaries, planters, soldiers, and tourists are partly on Hawaiʻi and partly on themselves, mobilizing phallic, racial, and property signifiers to know, to relate, and to seize. The new order written onto Hawaiʻi's land and people enables, normalizes, and legitimates the facts about the military with which this chapter began, recruiting Hawaiʻi into the supporting cast of the historical "play of dominations." But the expelled elements haunt the new order as its constitutive boundary, that which the hegemonic forces reject but cannot do without and thus can never fully obliterate. Those expunged but needed others also have voices in Hawaiʻi's militarized present—feminists, environmentalists, labor organizers, sovereignty activists, peace activists, artists and writers, and others with local allegiances and a critical eye for power—a cacophony that makes power more visible and more vulnerable to challenge.

2

Looking in the Mirror at Fort DeRussy

The Fort DeRussy Army Museum is a rich source of the military's understandings of itself and of Hawai'i. The museum site is itself an arresting inscription of the military presence in Hawai'i (fig. 6). It was once Battery Randolph, one of six low-silhouette steel-reinforced concrete emplacements housing fourteen-inch coastal artillery guns meant to protect O'ahu. The two huge cannons were never used in combat or much in practice because when they were fired "windows would shatter across Waikiki, crockery would leap off shelves, and tourists would whine" (Burlingame 1990: B-1). Obsolete since the beginning of the Second World War, the huge guns at Battery Randolph were dismantled and scrapped. The fortification itself could not be demolished, owing to its design to withstand direct artillery hits; demolition crews attempting to knock down the structure found that "the wrecking balls bounced off and shattered" (Burlingame 1990: B-1). Translating a problem into an opportunity, the army converted the structure into the present-day U.S. Army Museum. Trustees of the museum include retired and active-duty military officers, local political officials, corporate representatives, and Senator Daniel Inouye. The Senator's[1] fund-raising skills have led the trustees in an extremely successful capital improvements project, in which, in the words of then Executive Director Robert Thoads, "every penny of their donations is applied to storyline development" (quoted in Rho 1989: 11). And story-line development it is: the museum's

Figure 6. Coconut palms and military vehicles at Fort DeRussy Army Museum at Waikiki beach. Photo by Phyllis Turnbull.

narratives of wars, warriors, and citizens articulate the grammar and rhetoric of the U.S. military in Hawai'i.

Visited by 140,000 tourists annually, centrally located on Waikīkī beach, the army museum floats a complex enunciative economy of progress (with setbacks) and redemption. Starting with the legacies and relay points of an earlier discourse—the overwhelmingly gendered, colored, and classed narratives of "discovery," exploration, and salvation—the army museum segues readily into the origin story and the telos of war and the national security state. Feminized, the islands and the people were and continue to be reinscribed with meanings according to the needs of the eighteenth- and nineteenth-century colonists and the present-day military. The contemporary military does not have to create the discourse of darkness, lack, and feminine receptivity; it can simply ride on it. The prior production of the discourse by explorers, missionaries, and traders provides an already articulated set of opportunities for making statements and for keeping silent. Gendered colonial discourse can fade into the background, become the assumed backdrop of interactions without requiring any crude or blatant articulations, while remaining available as a streamlined and effective resource for maintaining Hawai'i's semiotic place in an order of domination and control. The army museum is replete with references to Hawai'i as an R-and-R center; smiling young "Polynesian-looking" women in hula garb embrace (usually white) soldiers with welcoming arms. While the most visible articulations of Hawai'i as a soft, welcoming, feminine place appear now in the strategic economies of tourism, the manly military continues to confirm itself and its eligibilities on a gendered horizon.

The process of the inscription of meaning was/is not a simple ink-on-paper act that is accomplished with a single swift stroke. Rather, it happens as a series of new normalizing processes take over and are both contested and entrenched: new categories, new names, new relationships, new histories. Historian and Director Tom Fairfull insists that the museum does not glorify war, and in many ways he is correct. Within the dominant rhetorical conventions and narrative strategies of militarism, Fort DeRussy's story is relatively wide-ranging and fair-minded: the enemy is not routinely demonized, civilians are not invisible, "we" are not uniformly virtuous. The political potency of the DeRussy story lies neither in its "bias" nor its accuracy, but in the seamlessness with which it produces and records a militarized understanding of Hawai'i and the world. The director intends for the museum, "like all Army museums," to be "used to support the education, training and recreation of Army personnel and as a community resource" (Rho 1989: 7). "We just want to show the way that it was," says Fairfull. "We wanted to make sure that what we were doing was educational, not propaganda, and not weapons displays geared to the 'button collectors and rivet counters,' as we call 'em" (Burlingame 1990: B-3). "The way things were" becomes a reified field of stable data, a coherent set of images that wait to be revealed. The narrative practices entailed in grasping those data, and the structures of consciousness required to fashion that coherency, are rendered invisible. Yet this effort is never fully successful; Hayden White reminds us that "even in the most chaste discursive prose, texts intended to represent 'things as they are' without rhetorical adornment or poetic imagery, there is always a failure of intention" (White 1978: 3). The politics of the heavily armored discourse at Fort DeRussy are not in the accuracy or inaccuracy of its facts, but in its constitution of the ground on which to decide what counts as a fact and what modes of comprehension to activate in understanding these facts (White 1978: 3–4). The Fort DeRussy story both reflects and produces an "institutionalized forgetfulness" (Michael J. Shapiro 1997: 22), which hardens the complex relations between the U.S. military, the U.S. government, and Hawai'i "into a hypostasis that blocks fresh perception" (White 1978: 3–4).

To unpack DeRussy's enunciative economies and deflate its currency requires attention to the operations of "style, figures of speech, setting, narrative devices, historical and social circumstances" (Said 1979: 21). Even the tightest narratives make their meaning claims by implicit reference to other, absent signifiers. Our goal is to track the deferrals that are entailed in "the interplay between presence and absence that produces meaning" (Moi 1985: 106). Taking militaristic discourses as "systems[s] of opportunities for making statements," we aim for a critical reading that can offer greater visibility both to the power that enables these statements and to the possibilities for making other kinds of statements and having them regarded as legitimate (Said 1979: 273). The dominant semiotic field has created Hawai'i as a discursive domain that invites and confirms the military's investments and self-understandings, as the obliging woman-native-other to its masculine, colonizing self, as opportunity to its actor, as blank page to its pen. A dislodging of this enunciative regime might create openings for a differently politicized set of understandings to make their appearance and do their work.

The practices required to reactivate the motility of discourse and make space for fresh perceptions call for careful attention to the dominant metaphors, recurrent figures of speech, authoritative grammars, and prevailing rhetorical moves that produce realist discourse. Fort DeRussy's narrative sports claims to The Big Is: "The Army, through its museum, attempts to present all facets of war in proper perspective" (Rho 1989: 8). The army's proper perspective does not blatantly hide so much as subtly disguise its inevitable incompleteness. Like the original cannons that were "mounted on a 'disappearing' carriage so that the recoil would cause them to drop behind the concrete wall after firing," the museum narratives ride on disappearing discursive practices that orchestrate and naturalize the museum's speeches and silences (Rho 1989: 7). A different sort of recoil can bring these implicit orthodoxies into the open, not to reveal a contrasting reality behind them but to enable contesting interpretive practices.

The structure of our reading of the museum narrative against the grain interrupts the structure of the museum itself. The former Battery

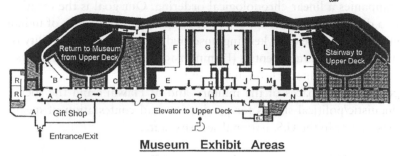

BATTERY RANDOLPH FLOOR PLAN

Museum Exhibit Areas

A.	Introduction Foyer	G.	Manning the Defenses	M.	Hawaii's Japanese
B.	Theater/Changing Gallery	H.	Army Aviation Takes Off		Americans
C.	Hawaiian Warfare	I.	The Winds of War	N.	Korea, A Limited War
D.	Camp McKinley	J.	December 8, 1941	O.	The Vietnam War
E.	Defending an Island	K.	Hawaii On Defense	P.	Gallery of Heroes
F.	Shell Magazine Replica	L.	Hawaii On Offense		

Figure 7. Floor plan of the Fort DeRussy Army Museum, the former Battery Randolph. Courtesy U.S. Army Museum.

Randolph is a single, long, and narrow corridor with galleries of various shapes and sizes opening off to one side (fig. 7). The military's story of Hawai'i, consisting mostly of photomurals, maps, and drawings with accompanying text, is arranged chronologically along the long spine as a linear narrative. Galleries of various shapes and sizes open off one side of the corridor; within these spaces particular aspects of the corridor's narrative are elaborated through presentations of material artifacts, written and spoken texts, dioramas, and more archival photomurals. Our own work, which aims at making space for other interpretations of the military's relations with Hawai'i, creates alternative dioramas of the rhetorical gestures that organize the museum's representational power: opposites, reversals, analogies, elisions, synecdoches, repetitions, and shifts in voice. These interactive tropes operate as linguistic gestures of insistence, producing and hiding (and at the same time, ironically, marking) preferred relationships between phenomena and necessary omissions within accounts. In the political work done by figures of speech, those smugglers of stolen goods hide in plain sight the spaces where other possibilities of interpretation have disappeared, and thus where they might reappear (de Man 1984: 201). Our tour through the Fort DeRussy Army Museum is organized to produce friction vis-à-vis the prevailing narrative, to resist the naturalization that typically accompanies a linear chronological ordering. Our goal is the creation of a different narrative space by "forcing the juxtaposition of unlike elements to the end of keeping unsettled and open the possibility of new spatial arrangements" (Dumm 1996: 41). It is a tour not so much of the military through time but of discourse in space, both the geographical spaces of the military and of Hawai'i and the social/linguistic/political spaces for production and contestation of meanings of war in the U.S. national security state.

THE GRAMMAR AND RHETORIC OF MILITARISM

Fort DeRussy sets out to tell the story of the military in Hawai'i and instead tells the story of the military's Hawai'i. The military is looking in the mirror at Fort DeRussy in that it looks at Hawai'i and sees itself, sees the Hawai'i it needs in order to be itself. The Fort DeRussy narratives construct Hawai'i—its origins, development, and destiny—by deploying grammatical and rhetorical devices that produce, disguise, yet simultaneously by their labor reveal the needed silences, the necessary yet unpalatable dependencies, the unarticulated alternatives. The grammar of the narrative establishes "the 'correct' or orderly use and placement of the constituent elements . . . the proper and acceptable situation and use of language" (Layoun 1992: 411). The narrative's rhetoric acts to convince us "of the efficacy and desirability of its terms and of the 'natural' relationship between those terms" (Layoun 1992: 411). Grammar, then, asserts the proper place of the elements of the story, while rhetoric works with those elements to persuade us of likely possibilities for interaction and resolution. Because there is both considerable overlap and potential divergence between the grammatical terms and the rhetorical gestures of a narrative, the two may both work closely together to establish a hegemonic story and also produce frictions that dislocate it.

Oppositions and Reversals

The military's pen(is)manship, in Hawai'i and elsewhere, writes its story with a distinctly masculine hand. It does its work through the establishment of dyadic grammatical terms, imposing the rule of two onto a field of many, followed immediately by the rhetorical assignment of hierarchical ordering: there are two kinds of forces in the world ("them" and "us," or, in the post–Cold War world, safety and danger), and one is benevolent while the other is suspicious. While the Fort DeRussy story permits a small amount of leakage here, in that sometimes "we" are incompetent and "they" are courageous, for the most part there is little space to figure places and persons that do not fit into either category, or that spill into both. There is little opportunity to articulate difference that is neither oppositional nor hierarchical, or equality that does not collapse into sameness, or proliferation that defies a linear rank ordering.

Construals of opposites allow the museum narrative to recruit everyday dyads to establish supporting patterns of similarity and difference and, importantly, to contribute to the naturalization of war as combat between two different sides. In a familiar contrast, Japan's defensive strategy is twice described as "fanatic," suggesting that U.S. defensive efforts were reasonable (early version of first gallery).[2] Similarly, Vietcong weapons are described as "diabolic," while U.S. troops are armed with more civilized "technologically sophisticated weapons" (Grunts vs. Charlie exhibit). A less predictable opposition occurred in an earlier display in the first gallery when the caption contrasted war with tourism: "In Hawaii alone, thousands of Japanese tourists arrive daily. Fifty years ago, it was very different." The contrast between peaceful, as opposed to bellicose, traffic across national boundaries doubtlessly reassured Japanese tourists that they are no longer the enemy, but the very insistence on difference, by putting the two events in close proximity, suggests similarity. For Native Hawaiians who see tourism as a different kind of assault on their indigenous ways of life, the juxtaposition could suggest that the two are not so different after all (Trask 1993). Similarly, a panel pronouncing "the amazing crosscultural influences shared by the US and Japan since the end of the Pacific War" reproduced a beer ad featuring a nearly life-size figure of a young Japanese woman, with large breasts, sitting atop several cases of Budweiser beer. She sported abbreviated denim shorts and cowboy boots. The sameness of consumerism overcame the difference of nationality: "This Bud's for you."

The construction of opposites enables reversals or inversions of familiar oppositions without fundamentally altering the terms of their relationships. Occasionally, the Fort DeRussy story reverses the usual equation of "us" with virtue and "them" with vice. Under the exhibit title Sins of Omission and Commission, U.S. military and civilian political leaders are taken to task for being unprepared for Japan's attack against Hawai'i. Perhaps the most overt expression of overconfidence came from Vice Admiral William S. Pye on December 6, 1941: "The Japanese will not go to war with the United States. We are too big, too powerful, and too strong." The navy limited its patroling to conserve fuel and lined up its battleships in neat rows, two by two. The army "considered sabotage to be the main threat; thus its planes were parked *wingtip to wingtip,* making them easier to guard" (emphasis in original). A general lack of alertness is criticized in all the branches of the service. Conversely, in the Pacific War gallery the enemy is described as brilliant: Admiral Yamamoto, planner of the Pearl Harbor attack, is credited as "Japan's master strategist," and Japanese pilots are recognized for their "great skill at low-level flying and accurate marksmanship with bombs and machine gun fire." This reversal exemplifies the determination of the museum's director to tell "the truth," even if it hurts. We made mistakes, the exhibit insists, and "they" were competent. This reversal, according to *Webster's,* establishes "an exception to the regular progression of a series of values." It does not contest the dualisms of us and them, good and evil, but merely temporarily causes them to move in an opposite direction. The net result is to lend credibility to the narrative, to establish it as education, an appropriate learning experience for soldiers and citizens rather than merely a bellicose display for button collectors and rivet counters.

This relatively minor reversal of competence and incompetence paves the way for the much grander telos of sacrifice and achievement, as represented in the final Gallery of Heroes. One portion consists of pictures or drawings of approximately thirty men from Hawai'i, with descriptions of the medals they were awarded. Another group of pictures and medals represents the six men from Hawai'i who received the Medal of Honor, all posthumously. Above each photo is a speaker emitting a recorded explanation of each man's heroic actions. The subdued voices of the recordings add to the somberness of the gallery. The forthrightness of the criticism of U.S. incompetence could be construed as a gesture of respect for those who died, a need to tell the story honestly so that the real heroes could be honored for their bravery and distinguished from those who did not properly do their jobs. While this is an understandable and honorable intention, the discursive outcome is to hold the two sets of

figures, the brave and the not-so-brave, heroes and incompetents/
villains, safely in place. Rather than questioning a man's heroic death
in war as the paramount masculine virtue, this outcome makes some
minor revisions in the distribution of heroism and irresponsibility.

A similar reversal operates within the crucial national security
dyad of safety and danger. The gallery Gathering Storm narrates
"the casual attitude of Hawaii's civilian population in the 30s." A
false sense of security was produced because "the powerful military
presence and thorough civilian defense preparations made most
people feel immune from attack." Photo subjects here include child
actress Shirley Temple, wearing an officer's cap, sitting on a cannon;[3]
the obligatory hula girls with leis; President Roosevelt dedicating Ala
Moana Beach Park; Babe Ruth posing with Duke Kahanamoku (a
famous local surfer and Olympic swimmer); a Pan-American Clipper
(Sikorsky's S-42 four-engine, pontoon-equipped airplane that pre-
saged the modern transpacific flights of an exponentially expanded
tourism). American soldiers' preoccupation with organized sports,
parades, and military routines is contrasted with the eminent danger
signaled by the Japanese war with China. "Local citizens, wealthy
tourists, celebrities, and even the President of the United States re-
laxed in beautiful Hawaii." The telos of the narrative suggests that,
even though the islands were heavily armed, they were not immune
from attack. It is, however, arguable that the islands were the objects
of attack *because* Pearl Harbor had become the home of America's
Pacific fleet rather than of the system of fishponds that flourished
earlier. Security and danger, posed against each other as opposites,
could well be reversed: the provisions for military security actually
created the danger by making Hawai'i a target.

Analogies and Elisions

Like opposites and reversals, analogies and elisions work together,
each enabling the other. Analogies foreground some resemblances or
proportions, while elisions nullify others. Both mobilize the available
grammatical terms with inclusive or exclusive rhetorical gestures. Together they establish or eliminate detectable similarities, recommend
for or suppress from consideration different possibilities for speech
and silence. Analogies between Hawai'i and Europe, or Hawai'i and
the mainland United States, establish relations of congruence between
otherwise unlike things. Analogies work as methods of coupling

diverse phenomena so as to render unfamiliar things familiar. Successful analogies encourage us to ignore the context of particular events or activities, to block out differences in order to equate otherwise unlike things. A corridor panel on the *ali'i,* the ruling chiefs of the Hawaiian kingdom, states, "Captain Cook found a feudal society in Hawaii, like Europe of the Middle Ages. Ali'i, powerful warrior chiefs, controlled the islands through heredity and kapu, a rigorous system of socio-religious rules." Hawai'i here is represented as similar to Europe, only lagging behind in the march of progress. Relations between *ali'i* and commoners are represented in purely military terms ("The ali'i required military service from the tenants on their land, and trained them regularly in the arts of war"), a very partial interpretation that leaves unstated the complex economic and religious affiliations between the classes. The rich culture of pre-contact Hawai'i is folded neatly into a military model: "Hawaiians sailed to their islands nearly a thousand years before Columbus' time, and developed military systems" (Hawaiian Warfare exhibit). While Hawai'i's history is militarized, Hawai'i itself is put "in context" by placing it in reference to mainland history. Events on the mainland serve as the anchor and frame for events in Hawai'i. A running header across the top of a series of exhibits reminds the viewer of what was going on in the "real world": "The U.S. army is born in Boston"; "Settlers move into Ohio, Indiana, Illinois"; "Gold discovered at Sutter's Mill, California." George Washington's inauguration in 1789 is coterminous with the rise to power of Kamehameha the Great.

Analogous links among importantly unlike things allow the museum's narrative to claim a seamless history in which early events just naturally evolved into their latter counterparts. The display titled Honolulu Fort: Symbol of Sovereignty relates:

> As a statement of independence for the Hawaiian monarchy, Kamehameha directed the construction of a fort to protect Honolulu Harbor and symbolize his sovereignty. Honolulu Fort was built of coral blocks, completed in 1817, and mounted forty cannons of various size, to deter foreigners, English, French, Russian, from attempting to seize control of the island.
>
> It remained a viable fort until 1857 and served as a military garrison, police station and a prison. Honolulu Fort foreshadowed the Coast Artillery of the U.S. Army in the defense of O'ahu.

In this analogy an early fort built by a Hawaiian monarch to deter foreign conquest is the natural predecessor of a later installation by a successful conquering army to secure its control. The irony, of course, is that the early fort clearly failed to deter (some) foreigners.

Other analogies at work in the DeRussy narrative include the familiar colonial relations of likeness between Hawai'i and pleasure, and between Hawai'i and children. In the Vietnam Gallery, Hawai'i is "remembered best as a Rest and Recuperation Center." A reconstruction of a bar in Waikīkī presents Primo Hawaiian beer among its wares, while a television displays a haunting video of news clips from the 1960s and 1970s. Antiwar demonstrators are juxtaposed with soldiers in trenches. Shots of Presidents Richard Nixon and Lyndon B. Johnson contrast with terrified Vietnamese civilians, while Led Zeppelin, The Beatles, Simon and Garfunkel, and other counter-cultural icons play in the background. "In just a few hours after leaving the steaming jungle behind," the display intones, "a G.I. was sucking up mai-tais in Waikiki's neon jungle." From behind this "incongruous taste of paradise" a hoarse, frightened soldier's voice periodically screams "Incoming!" The horror and danger of Vietnam are the backdrop to the safety and pleasure of Hawai'i.

Throughout the museum female hula dancers are shown entertaining tourists and soldiers. Yet the Go For Broke gallery announces that during World War II, "girls were in short supply." The innocent invocation of R and R and brief mention of "sleazy honky-tonks and hookers" elide the tight military and police control over prostitution in Hawai'i during the war and declines to document the vast administrative apparatus that was and is necessary for the regulation of female sexual labor near military bases (Anthony 1975: 23, 27, 40).[4]

It is estimated that there were around fifteen brothels concentrated in the few downtown blocks making up the red-light district. Each of the women who worked there is reported to have had sex with an average of a hundred men a day (Bailey and Farber 1992: 98, 100). Long lines of military men and civilians waiting to enter the brothels was a commonplace sight.[5] Other unregulated houses were scattered elsewhere throughout the city. Approximately 250 "entertainers" from twenty brothels were registered with the Honolulu police, a continuation of prewar policy (Allen 1950: 354; Anthony 1975: 40). Military control of prostitution under martial law was contested by the police; at one point prostitutes, upset because of the consequences of the feud between the two authorities, picketed both "the police station and the office of the military governor" (Anthony 1975: 40). The women who sold their sexual labor commanded significant leverage in the wartime seller's market, often making fifteen to twenty times the salaries available to other wage-earning women (Bailey and Farber 1992: 100). The coy reference in the DeRussy narrative to the short supply of girls both marks and disguises the tight organization of young women's social lives in Hawai'i, either by their anxious parents or by the restrictions of martial law, and glides over the military and civilian officials' importation and regulation of prostitutes on Hotel Street.

Similarly, Hawai'i as child to the world of warring adult nations works by analogy to proclaim a relation of likeness, this time between Hawai'i and an innocence that is endearing, if doomed. The exhibit on V-J Day proclaims: "Racial, cultural and economic barriers had been broken. Because of the war, Hawaii began to come of age." High wartime wages, increased tourism due to easier air transportation, expanded union organizing, proliferation of small businesses, and disintegration of the plantation system in favor of a more mobile labor force are all indicators that Hawai'i had "grown up." The ways of life that preceded this expanded participation in a hemispheric (if not yet global) political economy of wage labor and international tourism are relegated in this economic gesture to the quaint backwaters of political childhood. Inevitably and admirably, Hawai'i matured:

> World War II transformed distant exotic Hawaii from a fantasy into a familiar paradise for thousands of Americans, and demonstrated its importance to the United States. Hawaiians proved their patriotism and loyalty. *Statehood was close at hand.* (Aftermath and Beginning exhibit, emphasis in original)

History collapses into what Anne McClintock calls "a single, European Ur-narrative," "an organic process of upward growth, with the European as the apogee of progress" (McClintock 1995: 37). Differences that might have been recognized as equally valid options for social life instead become organized into a temporal hierarchy of early/primitive and late/developed. Hawai'i is marked not as socially different but as temporally prior "and thus irrevocably superannuated by history" (McClintock 1995: 40). Hawai'i's trajectory was already established, written in the telos of national identity that ironically disqualified Hawai'i's own sovereignty while folding Hawai'i neatly into the national security state.

Elision works overtime in the entrance to the long corridor of galleries, which offers the following introduction to the museum:

> Hawaii's military heritage is richly diverse. Military institutions, events, and technology have affected Hawaii's people since ancient times with political, social and economic impact.
>
> Our story tells of the men and machines which shaped that heritage; warriors who built a kingdom, soldiers who defended an island, citizens who served their country and sacrificed to keep it free.
>
> Hawaii's many ethnic groups share this proud heritage. Each has contributed in some way to the fabric of Hawaii's military past. This is their story, and the story of the U.S. Army in Hawaii. (Hawaii's Military History exhibit)

But the story that follows is difficult to read as "richly diverse"; instead, it appears as a predictably monotonous retelling of familiar Western fables of Progress, Sacrifice, and Redemption. The egalitarian reference to "Hawaii's many ethnic groups" elides the significantly different relations that Native Hawaiians and various immigrant peoples have maintained with the U.S. military and with each other.[6] The ethnic diversity and conflict in contemporary Hawai'i are absorbed into a unilinear story of proud warriors, brave soldiers, and freedom-loving citizens. A later exhibit titled Martial Law indirectly suggests some of these submerged conflicts:

> Just after noon on December 7th, [1941,] at the urging of the Army's Commander Lieutenant General Short, and with the concurrence of the President of the United States, Hawaii's Territorial Governor Poindexter proclaimed martial law: military officers moved into Iolani Palace and assumed all legislative, executive, and judicial powers.
>
> Martial law suspended constitutional rights, turned the civilian courts over to the military, imposed blackout and curfew, rationing of food and gasoline, censorship of mail and news media, temporary prohibition, realigned business hours, froze wages, and regulated currency.
>
> All civilians over 6 years of age were required to be fingerprinted. Except for taxes, General Orders issued by the Military Governor regulated every facet of civilian life, from traffic control to garbage collection. Violations were punished summarily by provost courts or

military tribunals; there was no right of appeal. Martial law remained
in effect for nearly 3 years, long after the immediate danger had
passed.

The minor self-criticism in the last two sentences is accompanied
by an unspoken racial subtext: why martial law here, and nowhere
else? And why was martial law continued for two years after the
pivotal Battle of Midway, which ended Japan's invasion threat (Fuchs
1961: 299)? Part of the answer lies in "the question of internal security weighed against the need for Japanese labor" (Okihiro 1991:
228). Constituting approximately one-third of Hawai'i's population,
and essential for sugar production, Americans of Japanese ancestry
in Hawai'i could not as readily be interned as those on the mainland
(Okihiro 1991: 195). Fourteen hundred Japanese Americans were
rounded up and interned at Honouliuli and at Sand Island or shipped
to the mainland camps (Menton and Tamura 1989: 149). Martial
law provided the apparatus for heightened regulation and surveillance of the remaining Japanese Americans, who were considered
both necessary and untrustworthy.

This abiding racial discrimination does not, however, completely
account for the lengthy imposition of martial law in Hawai'i. The
U.S. military had earlier threatened to take control of Hawai'i, in response to what the navy considered to be unacceptable racial turbulence, following the locally infamous Massie case. In 1931, Thalia
Massie, the wife of a navy lieutenant, alleged that she was beaten
and raped by five men of mixed Japanese, Chinese, and Hawaiian
ancestry. She gave conflicting testimony; the Honolulu Police Department was notably inept in its investigation; the prosecution case
was weak; the ethnically mixed jury was hung; Honolulu seethed
with excitement. Massie's husband and mother took justice into
their own hands. With two navy enlisted men, they kidnapped one
of the defendants and took him to Lieut. Massie's house where he
was shot and killed. His body was loaded into the back of a car
driven by Lieut. Massie. Accompanied by his mother-in-law and one
of the navymen, he headed for a cliff overlooking the ocean to dis-
pose of the body when the speeding car was stopped by a Honolulu
policeman.[7]

At their trial, all four were found guilty of kidnapping and mur-
der. Each was sentenced to ten years in prison. The navy, the domi-

nant service in Hawai'i at that time, had been outraged at the "failure" of the ethnically mixed jury in the first trial to convict the local
men. The verdict in the second trial prompted angry navy officials to
call in their biggest guns within the mainland political establishment
to gain control of Hawai'i. Dominated at that time by southern congressional leaders, the mainland forces would not accept the judgment of a local jury on a white southern woman. The commandant
of the Fourteenth Naval District in Hawai'i wired the Secretary of
the Navy, saying that "mixed blood juries were not capable of giving
justice to a white person" (Fuchs 1961: 189). In reply, the secretary
asked whether the fleet should be sent to Hawai'i. A bill introduced
into the Senate called for Hawai'i to be placed under naval commission government, a suggestion that had been floating around for several decades (Fuchs 1961: 189–91). Under such pressures, Governor
Lawrence M. Judd commuted the sentences of the four from ten
years to one hour. In his memoirs, Governor Judd said he was under
intense pressure from "an infuriated Congress" and that punitive
legislation was being considered. "Had I not acted as I did, I believe
that the form of government here might have been changed to some
form of government by commission" (Judd 1971: 168). The prisoners served their sentences of one hour in the governor's chambers
and immediately afterward left Hawai'i along with Thalia Massie.[8]

There is no mention in the Fort DeRussy Army Museum of the
repeated agitations by Congress or the navy to govern Hawai'i directly. The silence encompasses not only a pivotal case in local history but also a signal illustration of the U.S. military's sense of entitlement toward Hawai'i. Martial law takes on a different hue when
viewed, not as an exceptional response to overwhelming security
needs, or even as an excessive racial anxiety regarding the Japanese,
but as another stitch in an ongoing tapestry of arrogance and appropriation. A 6 A.M. to 6 P.M. blackout for civilians, eased to a dim out
by June 1942, was strictly enforced, although repair crews worked
twenty-four-hour shifts at Pearl Harbor, Hickam Field, Bellows Field,
and other bases, which were "illuminated like Christmas trees" (Anthony 1975: 59).[9] A further general order empowered the police to
confiscate houses and property if the inhabitants were "commiting a
nuisance or using the housing for immoral purposes or illegal purposes" (Brown 1945: 1). The military government's provost courts
required the guilty to buy U.S. war bonds, tried to force them to

donate blood, and assessed fines of arbitrary amounts, which the military then kept, although some of these kinds of fines normally made up part of the revenue base of the City and County of Honolulu (Anthony 1975: 46–60). The military wrote its own order onto Hawai'i, not the messy, even effeminate order of civilian government and rule of law, but the tough, no-nonsense order of hierarchy and command.

Other silences hide out in the open in the museum narrative. An earlier exhibit in the first gallery displayed a photo of Hiroshima in 1939, side by side with one of that city in 1985. In the older photo the city looked old-fashioned, with a single streetcar traversing the narrow streets of a deteriorating neighborhood. The contrasting picture portrayed a new improved Hiroshima, with wide modern streets traveled by plentiful new cars and buses. The omission of the bombing and its accompanying devastation made it look as though Hiroshima might have engaged in a vast urban renewal project. The grammar of the photographs' terms colluded with the rhetorical gesture of their juxtaposition across other silenced terms and gestures, as if to say that Hiroshima today looks much better than in 1939. Progress has been secured.

More elisions gesture toward a submerged story of contestation and conquest. The postcontact century of rule by the monarchy is characterized as "relatively peaceful," despite the assault on the indigenous culture and people by epidemics, missionaries, soldiers, traders, and businessmen (Hawaii's Military Forces exhibit). The eighth exhibit, Sugar and Soldiers: Reciprocity Treaty of 1876, claims that "King David Kalākaua granted the United States exclusive use of Pearl Harbor as a naval base and thus secured Congress' approval of reciprocity on August 15, 1876. Hawai'i and the United States were linked formally by military and economic issues." Struck from the record in these brief accounts is the passing of about two-thirds of Hawaiian lands into the hands of the sugar-growing foreigners who needed agricultural machinery and favorable tariff treatment for their crops. Gone also is the tumultuous election of 1874, which matched Queen Emma, the widow of Kamehameha IV and the great granddaughter of the younger brother of King Kamehameha the First, against David Kalākaua. The queen had closer ties to England than to the United States and was not as likely to cooperate with the U.S. military (Kuykendall 1966: 256). Kalākaua was declared king

Figure 8. In lieu of the overthrow: the main corridor presenting the chronological narrative of the military in Hawai'i in the Fort DeRussy Army Museum. Artist's rendering by Murray Turnbull.

after an unruly election. When Queen Emma's supporters rioted in protest at her loss, 150 marines from the USS *Tuscarora,* conveniently anchored in Honolulu Harbor, were sent to "restore order." The hurried, partial enunciation of "Hawaii's military heritage" conveniently compresses the story, contracting the disjunctions, expanding the continuities, making the story seamless, unilinear, and altogether reasonable.

Silence reigns in the museum displays with regard to the overthrow of the Hawaiian monarchy by Caucasian businessmen and the U.S. Marines. Neatly, an Exit sign and door mark the wall space that might, chronologically, have contained reference to the overthrow (fig. 8). The recent tape recording that provides commentary on the exhibits is a bit more forthcoming about the overthrow but sustains its own no-less-remarkable silences about its context and possible meanings:

> Continued growth of US business and government operations in Hawaii placed enormous pressure on Hawaii's monarchs to relinquish their control. Finally, in 1893 Queen Lili'uokalani was overthrown and a provisional government was created under the control of the citizens of Hawaii. (Tape at listening station 6)

The voice admits to the overthrow but configures it as a result of "continued growth," a seemingly natural and desirable process that

Figure 9. 'Iolani Palace draped in mourning black at the centennial of the overthrow, January 1993. Photo by Phyllis Turnbull.

"finally" had certain inevitable results. Further, the indirect voice identifies this "continued growth," which is after all what capitalist enterprises are supposed to do, as the source of "enormous pressure" on the monarchy. No agent can be identified and held responsible, no questionable land grab is visible, no explanation is needed for how that growth came to be or at whose expense. The queen "was overthrown," but no agent is detectable or responsible. In the implied democratic constitution of the provisional government by its "citizens," no mention is made of who counted as a citizen and who did not.

The 1993 centennial of the overthrow of Queen Lili'uokalani's government, organized by Hawaiian groups, filled in details missing from the dominant account. 'Iolani Palace, from where she reigned, was draped in black, and the American flag was flown at half mast (fig. 9). Plays and reenactments of events of the time told a story of a small handful of largely American businessmen, editors, bankers, sugar planters, and politicians who forced the queen out with the connivance of the U.S. Minister. Calling themselves the Committee of Safety, these men aborted the queen's intention to restore Hawaiian

Figure 10. American flag raised over ʻIolani Palace at the annexation of Hawaiʻæi, with U.S. troops and arms presiding. Courtesy Hawaiʻi State Archives.

control of the government through a new constitution. After the marines seized the government buildings, the committee proclaimed the end of the monarchy and cobbled together a government that was swiftly recognized by U.S. Minister John L. Stevens (Kent 1983: 62–63; Daws 1968: 270–76). The queen was imprisoned in her quarters and the provisional government came into existence. Five years later, America annexed Hawaiʻi (fig. 10). This telling of the story, reconstructed for the general public at the 1993 centennial events, dislocates the innocuous reference to "continued growth" with a compelling narrative of appropriation and loss.[10]

The museum's representations of Hawaiʻi enact a typical colonial reduction: Hawaiʻi's most important relations are those with its colonizers. Hawaiʻi is what the United States sees it to be: "critical to the military interests of the United States"; a part of the "Pacific Strategy"; "a crucial link to the United States' new possessions, Guam and the Philippines, ceded by Spain, and to the economic markets of

Asia" (Annexation exhibit). The viewer is reassured that the 1898 treaty of annexation was "offered by the government of the Republic of Hawai'i" (Annexation exhibit).[11] There is no mention of exactly who constituted the government of Hawai'i in 1898, of the tawdry history of the Committee of Safety, an early forerunner of the national security state.[12] Hawai'i is portrayed/produced as an obliging partner in U.S. military and economic expansion.

Synecdoche, Repetition, and Voice

Three further representational gestures that activate the Fort DeRussy narrative include synecdoche, repetition, and shifts in voice. In synecdoche parts come to stand for wholes, or wholes for parts, thus allowing the erasure of obstreperous parts and facilitating the selective repetition of privileged correspondences. In this way, inconvenient variety is erased from Hawai'i's political landscape. The interests of commercial agriculture in Hawai'i come to stand for all of Hawai'i, to *be* Hawai'i: "The American Civil War, 1861–1865, stimulated Hawaii's sugar industry. Reciprocity, duty-free export of sugar to the United States, became a goal for Hawaii." Accession of Pearl Harbor to the United States in the late 1800s in exchange for trade reciprocity is described as "mutually advantageous" (Sugar and Soldiers exhibit). No hint is given that the mutual advantage was shared by the sugar industry, whose products entered the U.S. market duty free, and by the U.S. military. Gone from view are the indigenous people and the immigrant workers who mainly peopled the islands and who worked in the cane fields. *Hawai'i* is constituted as a simple and unified term in the grammar of the story: *Hawai'i* now equals the owners of the sugar industry, "Hawaii's interests" are now theirs. In a later panel on the conclusion of World War II the display states that "statehood remained Hawaii's goal" (exhibit on V-J Day). The particular coalitions of businessmen, "reform"-minded newcomers, and the bourgeoisie who wanted statehood have become "Hawaii," while those Native Hawaiians and others who resisted mainland domination have disappeared.

Repetition of key phrases signals a space of significance in the Fort DeRussy narrative, a place where sustained iteration emphasizes an important lesson to be learned. The constant coupling of *surrender* with *unconditional* and *attack* with *sneak* carry needed reassurances to an American audience. That the surrender was unconditional assures us that we won it all, that no messy negotiations and embarrassing concessions marred the display of our primacy. We're Number One.[13] That the enemy attack was sneaky, as opposed to merely surprising or unanticipated, reassures Americans that the United States occupied the moral high ground. They sneaked up on us in the dark, while we stood up and fought like men. Such repetitions recite convictions without defending them or even articulating them fully; they work as rhetorical copy machines, reiterating an association again and again for sustained effect.

Periodic shifts in voice can accompany any of the aforementioned rhetorical gestures, massaging the narrative toward particular ends. According to *Webster's,* changes of voice work to advocate, equivocate, provoke, revoke, invoke. For a time the frequently changing first gallery contained an exhibit on the end of the Pacific battles during World War II. The orientation toward that war was aggressively defensive. Accompanying the enlarged black-and-white still photographs of action in the Pacific was a triumphant narrative in a somewhat hysterical register. In celebratory tones a caption declared, "On March 9th, 234 B-29s dumped 1,667 tons of incendiaries burning out nearly 16 square miles, and killing 83,000 people. Tokyo was devastated!" Liberal use of exclamation points punctuated the prose: "The war between North and South Korea may not be over!" (Did World War 2 Occur? display). The tone recited the excited declarative pitch usually associated with the screaming headlines in daily tabloids.

The narrative of a previous film on World War II naturalized war by rolling right over the bombs, the death, the civilian as well as military casualties by phrases such as "the most destructive war . . . a war of devastation and destruction." With war comes all those things, the film implies; they are just ordinary and expected. The abstractness of the language overcomes the particularity of the photos. The film presented the news coverage of Japan's surrender, including a parade down Beretania Street in Honolulu. The parade included numerous military marching units, a USO float with a hula dancer in a grass skirt, and a Chinese dragon accompanying representatives of "the Chinese community." The current audiovisual presentation titled The Military Heritage of Hawaii features Senator Inouye as narrator: "When Western civilization first came to Hawai'i, our country was in the midst of a Revolutionary War. . . . Honolulu again felt the tramp of military boots." This film combines paternalistic descriptions of the "generous and benevolent hand" of the "patient and peace-loving monarchy" with oblique language that hides agency behind the heavy hand of History. The violence of Annexation is rewritten as a friendly handshake: "The United States solidified its ties with its mid-Pacific neighbor." In 1950 "word came that freedom was again threatened," bringing Hawai'i into the Korean War. The Vietnam War "once more pressed Hawai'i into service" as an R and R center, a home for military families, and a launching point for troops. The people and institutions who spoke these words pressed this service are rendered invisible and unaccountable. They have disappeared into the rhetorical magic of patriotic duty and national need. The film concludes that the actions of "our fighting men and women . . . make our country great," with no sense that the first person, plural pronoun might operate as a grammar of conquest.

Back in the main corridor of the museum, the text accompanying Hawaii's Military Forces: Soldiers of the Crown seeks refuge in an indirect voice:

> The monarchy maintained a Royal Household Guard of infantry, cavalry, and artillery, whose function was mainly ceremonial. As the years passed and rivalries developed, paramilitary organizations, such as the Honolulu Rifles, were organized as the business community's counter to the King's forces.

Questions such as why rivalries developed, what they were about, who the business community was, or why a business community would organize a military force against the government all go unaddressed. Rivalries simply "developed" as the years passed; no agency is visible in charting such rivalries; like Topsy, "they just growed." In the display of map and text titled The Great Pacific War, war itself is presented as the agent; it has needs, raises questions, makes demands. Similarly, the exhibit Aftermath and Beginning: Nuclear Deterrence and Limited War, evades any attribution of agency: "Atomic weapons brought an end to World War II." The use of such indirect voice allows the narrative to avoid the attribution of responsibility that direct voice would undeniably suggest.[14]

The museum's representational practices produce a certain piety toward war and the national security state. They do the grammatical and rhetorical work necessary to keep the narrative on track, to ar-

ticulate the questions and contain the inquiry within an implicit discursive framework assuring the legitimacy of the state, the solemn necessity of war, the inevitability of Progress. Our tropical tour, in contrast, aims to articulate the narrative's hidden dependencies on these
linguistic practices, to take them out of the arena of "common sense"
and into one of contestability, to make them one way, rather than the
only way, of telling the story. We are looking for "the virtual fractures
which open up the space of freedom" so that other accounts of war,
the state, and Hawai'i can be heard (Foucault 1988: 36).

BRINGING ORDER TO ISLANDS

The story of the military and Hawai'i that the museum relates is a
challenging one. Its representations must continually normalize the
military's presence, investing it with an ontological givenness. In
light of the necessary partiality of representation itself, such fixity is
provisional at all times. At DeRussy, several displays in particular
threaten to undo such stability: the support given by businessmen to
an insurrection against a legitimate government; the maintenance of
a suspiciously long period of martial law; the organization of female
sexual labor while civilian "ethics" are regulated; the rendition of
Pearl Harbor as a sign of security, not an attractive target; the
graphic video images from Vietnam juxtaposed with pious declaration that "the disengagement was honorable" (Vietnam gallery).
Another near miss occurs in the portrayal of the Army Corps of Engineers' practice of writing order onto the land at the turn of the century: mapping the island, laying out roads, dredging harbors, building lighthouses, and constructing housing for army troops (Corps of
Engineers; Construction Support exhibit): "Many of Oahu's roads
were constructed or improved to speed the movement of troops"
(Repelling Invasion; Island Strategy exhibit). "Because of the space
available and its central location in Leilehua plain, Schofield Barracks
was developed to house the majority of the island's mobile defense
forces" (Housing the Troops; Schofield Barracks exhibit).[15] This naturalizing story continues the use of passive voice to render the mod-
ernization of O'ahu unremarkable, while refraining from attribu-
tions of agency. O'ahu's roads were simply "developed," as though
nothing had been there before, as though the land were empty until
the military did something useful with it. Schofield Barracks just
naturally grew up on Leilehua plain, where space was "available."

Nothing had to be seized, no one displaced, no previous writing of order by the indigenous people was erased. Oʻahu was a blank slate waiting for the military to write order upon it.

The Army Corps of Engineers' narrative also foregrounds the trope of "islands" to replace, rhetorically, accounts of politics with apparently simple "facts" about "nature." On the second floor of the museum, the Corps continues to write the military presence and order onto Hawaiʻi—and other Pacific islands. The theme of the Corps' Visitor Center's exhibition, People, Islands and Water, situates the Corps in the Pacific with photos of Pacific Island people, tropical plants, waterfalls, and sunsets. The foyer is paneled in Hawaiian koa, and as one enters a recorded voice says, "E Komo Mai" (Come in). A slick multimedia production, People, Islands and Water depicts life in the Hawaiian Islands and in the South and Western Pacific Islands. The narrative implicitly participates in the hegemonic view of the Pacific, one that attributes the essence of islands to a simple spatial "truth" about them: the ocean is big and empty, and real estate is scarce. Gavan Daws articulates this view in his remark that "the scale of the ocean—70,000,000 square miles—reduces even the biggest of these island groups to insignificance. That is the geographic reality of the Pacific" (Daws 1980: xi). The materiality of water and land is confused with the western interpretation of this water and land, so that "significance" and "reality" are taken to be self-evident facts rather than historical, and contestable, accomplishments. Another telling of the water and the land produces another kind of order. To Epeli Hauʻofa, there is "a gulf of difference between viewing the Pacific as 'islands in a far sea' and as a 'sea of islands'" (Hauʻofa 1994: 152). The first envisions small and isolated bits of land remote from places that matter where the second stresses the dynamic relations among humans, water, and land. To the peoples of Oceania, who have lived in the Pacific for more than two thousand years,

> their universe comprised not only land surfaces, but the surrounding ocean as far as they could traverse and exploit it, the underworld with its fire-controlling and earth-shaking denizens, and the heavens above with their hierarchies of powerful gods and named stars and constellations that people could count on to guide their ways across the seas. Their world was anything but tiny. (Hauʻofa 1994: 152)

In the DeRussy account "islands" serve as "the governing metaphor of [the] historical account"; they act as a *"heuristic device which self-consciously eliminates certain kinds of data from consideration as evidence"* (White 1978: 46, emphasis in original). "Islands" segue from Hawai'i to the Pacific Ocean, and they enter this narrative as dreamwork, trailing no clouds of political history, social institutions, or differentiation. The Corps' narrative intones: "In the eye of the mind, islands are idyllic. . . . Truly, islands are dreams." The straightforward narrative of the show is that, despite their beauty and their "South Pacific" setting, islands need reworking by the Corps to make them physically safe for daily life, economically productive, and militarily secure. The narrative device of the exhibit disguises the agency behind this claim for reworking by making the islands themselves into the agent: "Islands need . . ."; "islands require. . . ." The "daily life" on O'ahu whose safety is being assured is, implicitly, a corporate daily life, which includes six million tourists per year, a chronic housing shortage, declining possibilities for agriculture, and limited land area, much of it marked for military use. It is also the daily life of a national security installation. In the remarkable grammar of the Army Corps of Engineers, *islands* operates synecdochically to give birth to the Pacific that the military wants to see.

Hidden in this essentializing rhetoric of "islands" are the specific histories of their repeated colonization beginning several centuries ago. The subsequent roles of the islands include "hosting" the ferocious battles of the Pacific theater of World War II, and the later "nuclear colonization" by Britain, France, and the United States. "Islands" came under the control of the United States as a result of war (the Philippines and Guam after the Spanish-American War, and Micronesia after World War II) or other forms of violence (Hawai'i after the overthrow of its government, and Sāmoa following a contretemps between Germany, England, and the United States). "Islands" have different geological histories as well. They vary by size, fertility, topography, climate, and relation to the open ocean. Some are atolls surrounding lagoons; some are singles, buffered by reefs; others face the open ocean unprotected. Still others may soon be underwater altogether if rising sea levels, driven by global warming, continue (Roylance 1997: A-24). The people living on several islands depicted in the museum are both Polynesian and Micronesian; they are culturally and linguistically diverse, with disparate traditions; they have diverse relations to the land, whether those of "class, family, religion [or] myth" (Leibowitz 1989: 86). Most of the Pacific islands are poor—not "naturally" so but because repeated colonization has disrupted their indigenous means of subsistence, diet, social relations, and relations to the land. The introduction of a money-based economy was accompanied by a dependence on expensive imported goods, a decline in health and living standards, and a horrendous shock to indigenous ways of life (Alexander 1984; Ishtar 1994).

After the Second World War, in a strategic reversal of "needs," islands, especially in Micronesia, became "needs" for the United States. As Arnold Leibowitz mildly puts it, "military doctrine saw the islands of [Micronesia] as an important forward base position for the United States, on the one hand; and, on the other, as dangerous to the U.S. mainland if occupied by a large, hostile power" (1989: 4). It was surely a compelling need; how else to account for America's taking possession of two thousand islands (most uninhabited by humans) and the surrounding ocean, an area roughly equivalent to that of the continental United States? It was a long-range need anticipated as early as the 1943 Cairo Allied summit meeting. Glenn Alcalay reports that in 1945, the *New York Times* lobbied for permanent military bases in Micronesia on the grounds that the islands were as important to the United States as Hawai'i (Alcalay 1984: 27).

A macabre doubling of the "need" of islands themselves emerged in 1946, in those early days of the Cold War weight-lifting contests between the Soviet Union and the United States. In a match made more in a strategic planning session than in heaven, the "needs" of certain Micronesian islands coincided with American military desire. The islands became "natural" sites and targets for U.S. postwar atmospheric atomic testing and missile experimentation. Coupled with the need for some islands to be targets was the need for some to be depopulated.[16]

In 1946 the first of the 106 atomic tests was carried out in the Pacific by the United States (Davis 1996: 8–9). An atomic bomb was exploded over a fleet of ninety-three World War II ships anchored in Bikini Atoll in the Marshall Islands. Operation Crossroads symbolically threatened the Russians with the destructive capacity of American atomic weapons. It was also, according to Jonathan Weisgall

(1994, 1996), an operation about whose outcome its agents were unsure. Crossroads represented intense interservice rivalry over control of the atomic bomb, which, because it could be dropped from airplanes, potentially might make a large fleet unnecessary. It equally represented, again according to Weisgall, the arrogance and ignorance of the testing program, and a secrecy that fed the arrogance and excused the ignorance. (At this time, alleges Weisgall, even President Truman did not know how many atomic bombs the United States possessed.) The threat to the Marshallese was less symbolic, more immediate, and longer lasting. Bikini Atoll had been selected for the tests by the military because it was isolated from shipping lanes and population centers. Military planners believed that prevailing winds would carry any fallout even farther away. The atoll provided a superb natural harbor to accommodate the fleet that was soon to become neither fleet nor a fleet. The strategic planners shared the imperial gaze with the discoverers of earlier centuries in seeing a convenient emptiness of the Pacific. Equally important, Bikini was within reach by the bomber designated for the test. And, finally, its population of 166 made it politically insignificant (Alcalay 1984: 27).[17] The Bikinians, converted Christians, agreed to be removed when an American naval officer, appearing after Sunday church services, compared them to the children of Israel who had been led to the Promised Land. They were assured that the forthcoming test would result in "kindness and benefit [for] all mankind" (Alcalay 1984: 28).[18] The social dimensions of such a removal (their relations to the land, the spatial geography of their way of life, their ability to nourish themselves physically and spiritually—all the features, in short, that give agency to a people) were minimalized in this opening move of the nuclear colonization of the Pacific.

The second bomb was detonated underwater with far more destructive force. Confirming the unboundedness of self-authorized desire, at least one island was simply atomized. Films of the explosion show a large ship turning turtle; the first wave produced was ninety-four feet high. Most lethal of all was a radioactive rain that completely contaminated the target fleet, the entire lagoon of Bikini, and the surrounding islands. This unanticipated result led to the cancellation of a proposed third test (Weisgall 1996).

Meanwhile, back at the previously uninhabited island (Rongerik) to which the Bikinians were moved as the first station on their hegira (which continues to this day), material conditions were busily undoing the essentializing rhetoric of "islands": "Rongerik's resources had been greatly overestimated and were, in fact, inadequate" (Kiste 1974: 78). As Alcalay puts it, "Rongerik was uninhabited because it was uninhabitable" (1984: 28). Essential elements of their diet, such as fish, coconuts, arrowroot, and pandanus, contained toxins. Alcalay also reports that the navy relocation officer, sent to check on their health, urged them to develop more self-reliance (1984: 29). Eventually, after a year of near starvation and neglect, most were moved temporarily to Kwajalein and, finally, to the island of Kili. Kili is tiny and desolate, its soil poor. Open to the ocean, the island lacks a lagoon, which prevented the kind of fishing that was traditional with the Bikinians. Following this move, the naval officer serving as deputy High Commissioner of the Trust Territory observed that if "the Bikinians do not adjust to Kili, the United States will have a new headache. *The Navy is running out of deserted islands* on which to settle these unwitting, and perhaps unwilling, nomads of the atomic age" (quoted in Alcalay 1984: 30, emphasis added).

As the "needs" of islands become doubled back on themselves, so does the agency of islanders in this statement. As objects of security discourse, they are removed from their homeland like goods set out on the curb for removal by a municipal refuse service, but when relocated as subjects in an alien land, they fail the test of initiative and an appropriate work ethic. When it was judged, on the basis of a 1968 survey by the Atomic Energy Commission, that Bikinians could return to their island, planning, replanting, and construction commenced; by 1972 a number of Bikinians had returned. They were abruptly removed again in 1978 after tests showed that they may have ingested more radioactive material than any known human population (at that time) (Weisgall 1996).[19] Today, 2,200 Bikinians continue to live on Kili (Wright 1996b: B-1).[20]

Eight years after Crossroads, a hydrogen bomb a thousand times more powerful than the Hiroshima atomic bomb was exploded farther north in the Bikini Atoll. The early weather forecast for Operation Bravo predicted a north wind, but it became easterly before the detonation. American ships to the east were directed to move south, but the people living on the eastern islands of Rongelap and Utirik had no warning. Several hours after the explosion, radioactive rain and/or powder fell on them, as well as on the crew of a Japanese

fishing boat, *The Lucky Dragon,* exposing them to the same radioactive dosage that persons living less than two miles from the explosion at Hiroshima would have received (Weisgall 1994: 32; 1996).
Twenty-eight servicemen (attached to a weather observation station
on Rongerik) and 236 islanders were diagnosed with radiation sick-
ness and afflictions caused by their exposure (Davis 1996: 8).[21] It
was three days before the badly injured and sick people on Rongelap
and Utirik were picked up by U.S. forces and taken to Kwajalein. Because of the radioactive fallout, the two islands will be uninhabitable
for thirty thousand years (Seager 1993: 64). As for the people, fetal
death rates spiked to 33 percent immediately after the blast and remain high. Joni Seager reports that of those who were younger than
ten at the time of the blast, 52 percent had developed thyroid cancer
or abnormalities by 1966, a rate that increased to 69 percent by
1976. Genetic disorders, cataracts, respiratory diseases, stillbirths,
and numerous kinds of tumors are still present among the population (Seager 1993: 64; see also Lutz 1984).

American operations directing nuclear testing in the Pacific were
based on Kwajalein, the largest atoll in the world. Now known as
the Kwajalein Missile Range, the atoll is the splashdown point of
intercontinental ballistic missiles fired from California, and a site
used for conducting Star Wars research (Alexander 1984: 13; Wilson
1995: 20–21). To develop the missile range, the military removed
Marshallese residents of Kwajalein to tiny Ebeye Island, with the
promise of food, housing, electricity, and the other amenities of a
small city. The few Marshallese who now work at Kwajalein and
commute from Ebeye daily confront the American military presence
and a way of life that contrasts starkly with their own. Ebeye's sixty-
six acres is home to nine thousand Marshallese living in extremely
crowded barracks with little infrastructure. They buy and eat expensive imported processed food and suffer high rates of malnutrition,
infant mortality, and unemployment. Their access to the lagoon (and
its fish) is restricted, as well as their access to other islands in the atoll
on which they raised or could raise food (Seager 1993: 63; Ishtar
1994: 26–27). Their former ancestral home, forty-three hundred
miles from Los Angeles, has become a temporary duty post for five
thousand Americans, who live in homes with trees, grass, streets, a
hospital, a nine-hole (but "hard to par") golf course, and signs reading, "No Marshallese allowed on these premises" (Horwitz 1991).[22]

The breathtaking ferocity with which the new order was written on the Pacific was and remains frightening, difficult to encompass or comprehend, but nevertheless compels analysis. The intertwined narratives sequestered within the myth of the virgin lands assist in unraveling the violence. In the narrative of the patriarchy, to be virgin is to lack both desire and agency, to await passively "the thrusting male insemination of history, language and reason" (McClintock 1995: 30). In the felicitously twinned colonial narrative, colonized people cannot "claim aboriginal territorial rights" to virgin land because the indigenous people are outside the historical discourse that endows such rights. The strategic formation combining patriarchy and colonialism assures the proper "sexual and military insemination of an interior void" (McClintock 1995: 30). The French military, in its nuclear testing program at Moruroa (Mururoa), gave a woman's name to each of the craters they gouged out of the atoll (Goodyear-Kaʻōpua 1997: 8), suggesting a violent affirmation of the old Baconian metaphor of raping the land.[23]

Indigenous understandings of the land also gender it feminine, but with very different consequences. Rather than an empty place waiting to be filled, land for some Polynesians is seen as a powerful and sacred space that participates in the creation of life (Goodyear-Kaʻōpua 1997: 9, 12). For the Maohi people of what they call French-Occupied Polynesia, women and the land, coded feminine, are respected for their valued role in creating life. In Jennifer Noelani Goodyear-Kaʻōpua's words, "Traditional Polynesian culture recognizes the centrality and power of procreation. This power can be coded both feminine and masculine. When female and male elements come together, life is born" (1997: 12).

In place of a story of mutual regard, the myth of the virgin land declared open season on indigenous lands. The Articles of Confederation granted the national government "broad authority to acquire and govern the Western lands"; the Land Ordinance of 1785 established the method by which the empty or emptied lands were to be surveyed and divided.[24] The Northwest Ordinance of 1787 articulated the underlying goal of territorial acquisition as eventual statehood.[25] McClintock is not straining in her analysis of the myth: in its patriarchal dimension the western lands were indeed empty and waiting (for those who emptied them). In the myth's colonial register, colonized people cannot make aboriginal land claims to virgin or empty land. Both stories disavow agency by women and natives, assuring gendered military authority over vacant places and obsolete people. For the continental territories, statehood was the telos toward which they moved. But in its consideration of the constitutional status of off-shore territories in the insular cases, the Supreme Court was reluctant to extend this form of political order.[26] Self-governance proved, in judicial eyes, to be a practice best reserved for the continental land mass and not necessarily appropriate for those areas not "integral parts" of the United States (Leibowitz 1989: 23). (This view was clearly abandoned selectively on two later occasions: Alaska and Hawai'i.)

The American capture of Micronesia from Japan at the end of World War II released anew both the sexual and racial narratives of the empty land, further confirming the Pacific Islands' unfitness for self-rule. The violence that vaporized an island at Bikini, blew one island into two at Enewetak (Eniwetok), irradiated and simply destroyed a great deal of Micronesia's "scarcest resource—land" (McHenry 1975: 59), played after the war in a different political register. Most of the land that was not destroyed was seized, confirming McClintock's rapacious metaphor of the "sexual and military insemination of an interior void" (McClintock 1995: 30). At U.S. insistence, Micronesia was defined by the United Nations in 1947 as a "strategic" Trust Territory (the only one of the eleven trust territories created and supervised by the UN after World War II) (McHenry 1975: 34); in one fell swoop Micronesia was rendered perpetually dependent (of course, with lip service paid to "development") and was configured primarily in relation to U.S. military interests.[27] As Donald McHenry mildly remarks, such a designation "has allowed the United States to exercise virtually complete control over the territory" (6).

Changing "strategic" American interests have continued to anchor the U.S. desire for permanent access to Micronesian land while also continuing to play to the colonial narrative that disavows the agency of the colonized.[28] Roxanne Doty suggests that classificatory schemes are ways of creating hierarchies among humans; once naturalized, the categories become at-a-glance referents that reproduce the structure of dominance (Doty 1996b: 10). Micronesia, to the hegemonic interests of the U.S. security state, is simply and obviously a place to which it is entitled. Out of step with History, the entire Pacific is seen in colonial eyes as a place where, as Victor Hugo's metaphor suggests, God's footstep had never trod. So of course the end of one colonial era is simply taken as a welcome opportunity to begin another, rather than perhaps a chance for a struggling, embattled, but persistent way of life to make its own way.

In the early days of the American trusteeship of Micronesia, the navy hired anthropologists to "conduct research throughout Micronesia to gain increased insight into local political and social structures" (Wilson 1995: 4). The role for "house anthropologists" was later expanded to aid U.S. officials in the implementation of military and political policies.[29] Their political goal was to effect some "permanent affiliation" between the islands and the United States, while hoping to avoid independence (McHenry 1975: 18). The U.S. military sought unqualified access to the lands and harbors of Micronesia; U.S. political strategies worked to secure this military goal. American-sponsored "development" fractured Micronesia into four entities—the Commonwealth of Northern Mariana Islands, the Federated States of Micronesia, the Republic of the Marshall Islands, and the Republic of Belau (Palau)—with four corresponding sets of local elites increasingly dependent on U.S. largesse. American responses differed as each group of islands articulated its particular preferences for a posttrusteeship status. But always, the desire for access to land anchored U.S. strategies, and the classificatory system of nations and territories legitimated U.S. entitlements. "Trust" territories are children in the family of nations.

Belau's struggle for agency gives a clear measure of U.S. desire. The United States had proposed a new Compact of Free Association, which would have given it free access to Belau's deep water port of Koror, to several thousand acres of land restricted to exclusive military use, and to thirty thousand other acres of land for military maneuvers. Meanwhile, the people of Belau held a constitutional convention in 1979 that proposed an independent democratic government with strict prohibitions on the foreign ownership of land and the storage of nuclear weapons. A final provision of this widely supported constitution required a 75 percent affirmative vote by the citizens of Belau to enact constitutional changes. American efforts to overturn, nullify, and change these provisions were thwarted for nearly two decades by the vigorous and astute political actions of Belauans. Opponents of the U.S. goals included a group of women elders (although earlier "house" anthropologists had discounted women's political agency), who pursued their defense of their country to the United Nations, the U.S. Senate, and to world opinion generally. The United States refused to renegotiate the compact but was not, until 1994, able to muster the required 75 percent of the vote to amend the constitution. As elsewhere in Micronesia, American policy had been to increase revenues sent to local political leaders to build support for the American case. Rarely, if at all, did the money provide the basis or stimulation for economic growth. Much of it went into a proliferation of government services and jobs, ultimately giving local political leaders the needed leverage to change the constitution. In Belau this leverage came after the murders of several political supporters of Belau's constitution, and after deep fissures split the extended kinship groups that wove the social fabric of Belau. When the Compact of Free Association with the United States was signed in 1994, the American military gained access to any Belauan (Palauan) land within sixty days of notification.[30]

In other parts of Micronesia and the Pacific, the military desire has similarly been met, for the time at least. The U.S. military now controls a third of the land on Guam. In the 1980s a proposal to replace colonial with commonwealth status was approved by Guam's voters, but it gathers dust in Washington while the Pentagon warns that "greater local authority would impair the military's ability to react in a crisis" (Kristof 1996: 14). The Commonwealth of Northern Mariana Islands granted rights to a major portion of the island of Tinian and conditional access to parts of Rota and Saipan. Under the compact with the Marshall Islands, the American military won continued and exclusive use of Kwajalein and, in addition, conditional access to other islands.

None of these complex geographies and colonial histories is visible in the Army Corps of Engineers' narrative, where the governing metaphor of "islands" gives directions for finding a different set of associations—not the specifics of political, economic, and nuclear colonization and its aftermath, but the technologically proficient categories of resource management, harbor development, shoreline protection, and flood control. The ongoing hunger for military access to "islands" continues the feminization of the Pacific islands by the manly military. The rhetorical reversal that attributes the desire for colonization to the "islands" themselves suggests both a feminine invitation and a sexualized vulnerability. "Islands" are incomplete; they lack. They are both damsels in distress and seductive females. Yet there is an unavoidable excess in this account. The narrative of the Army Corps of Engineers inadvertently recruits Hawai'i into the context of Pacific islands, rather than simply as part of the United States of America. The grammar of "islands" invites the viewer to think of Hawai'i not as the fiftieth state, sort of like California but farther into the ocean, but as more like Belau or the Marshalls or Bikini Atoll. The rhetoric of needs and wants that are attributed to "islands" operates in the military narrative to naturalize and disguise colonization; but it could also inadvertently invite one to imagine "islands" as needing and wanting other arrangements—independence, for example, or environmental safety, or perhaps justice.

3

Constructing and Contesting the Frame at Fort DeRussy

The military's raison d'être in Hawai'i (and elsewhere) is to guarantee national security, that is, the military exists in order to fight and win wars. Yet these two figurations of the military's mission—waging war successfully versus defending national security—are not, after all, the same. There is both interdependence and incompatibility between the discourse of war and the discourse of national security as ways of representing the military's legitimacy. The frictions and the attachments between the two can be seen in the different actions, metaphors, narratives, loci of responsibility—contrasting world constructions, in fact—that are associated with each. Edward Said notes that a mix of powerful discourses can interact to produce a discursive "strategic formation," a dense weave of texts that enables hegemonic accounts to "acquire mass, density, and referential power among themselves and thereafter in the culture at large" (Said 1979: 20). The interactive discursive economies of war talk and national security discourse at Fort DeRussy make Hawai'i into a coherent militarized subject matter and put into circulation a "common currency" through which this coherence can be produced and maintained (Said 1979: 213).

The relation between the two discursive realms can be thought of as an example of Roxanne Lynn Doty's concept of "double-writing."[1] Doty characterizes double-writing as the "successful practices . . . that produce the state's powerful image and simultaneously conceal this production" (1996a: 176). Following Homi K. Bhabha, she links the production of the state's powerful image with the pedagogical aspects of language, while connecting the concealment of that production with language's performative dimensions:

> The pedagogical refers to the givenness of an object, the presumption of its self-evident presence and essential identity (e.g., the state, the domestic community, the inside). . . . The pedagogical is intervened in and contested by the performative (the productive aspect of language). The performative introduces a "gap" or "emptiness" to the signifier by casting a shadow between it and the signified. (1996a: 176)

In the "double-time" of the pedagogical and the performative the production of claims to timelessness is hidden in plain sight (Bhabha 1990: 297). The pedagogical gives discourse "an authority that is based on the pre-given," while the "repeating and reproductive process" of the performative marks "a process of signification that must erase any prior or originating presence" (Bhabha 1990: 297). The interactions of pedagogical and performative operations produce a strategic formation that both anchors "Hawai'i" in the web of the familiar and offers some hints about the making of anchors.

Within both war talk and national security discourse important tensions operate. Each depends on assertions about what can be taken for granted in the world, and each disguises those assertions in order to maintain its potentially insecure foundations. Before investigating further this level of double-writing *within* the discourses, we want to map another level of doubleness, that which characterizes the relations *between* the two discursive practices. Our analysis, then, focuses on a kind of doubled double-writing. That is, each way of configuring international bellicosity simultaneously confounds and underwrites the other, while both call on parallel internal dependencies/ interruptions between the given and the manufactured.

Looking first between the two discursive fields, the contrast between war talk and national security discourse reflects the incommensurate social registers within which they function. War's companions are heroes and victories, to be sure, but also destruction, loss, death, pain, dirt, disease, hunger. The world of war is one of countries—our nation and others, some "friendly," some "hostile"—most armed and capable of waging a war that we might not win. War carries the risk that we might lose a way of life, and life itself.

The seductive glory of war talk, of sacrifice and redemption, has embedded within it the implicit threat of bereavement, suffering, and annihilation. The vast formation of graves and the pious chapel at Punchbowl National Cemetery (see chapter 4) are expressions of war talk, its excitements as well as its regrets. One of the weaknesses of war talk as an ongoing legitimation of the militarization of the world is that it too readily generates its opposite, peace talk. While there are available strategies within war talk to make war sound at least temporarily attractive—duty, honor, glory, manhood, risk, excitement, victory, destiny—there is also an undeniable linguistic affirmation in war's opposite, peace. While peace may be boringly effeminate, it connects readily to prosperity and serves as war's justification: as one army recruiter told one of our researchers in the midst of the Gulf War, "We're about peace, not war."[2] At the same time, the limits of pacifism are suggested by its close links to the universe of war; the opposite of war talk, peace talk cannot counter the differently ordered insistencies of national security discourse.

National security discourse lowers a different stage set, one of a powerful sovereign state whose continued security "compels it to coordinate and discipline . . . the peacelessness that reigns without" (Klein 1989: 101). Through crisis management, intelligence gathering, and maintenance of a "high state of readiness," a technologically sophisticated weaponry can be aimed at the world's depersonalized or demonized forces of disorder. National security discourse deploys a particular kind of knowledge, a tactical view of the world, a problematic of "strategic control at deep levels" (Haraway 1989: 110). This way of talking about military objectives and entitlements recruits the somewhat more sedate, rationalist language of stability, predictability, and development. The world is configured not so much as a vast potential battlefield under the rocket's red glare, but more as a calculable reserve of population and resources, always potentially unstable yet amenable to strategic reasoning.

The extensive mosaic maps, charts, and texts in the Punchbowl Memorial are expressive of the thoroughness and rationality of national security discourse. So is the analysis offered by Admiral Charles Larson, former Commander-in-Chief of the U.S. Pacific Command (fig. 11), which he called "the largest unified command in the world" (United States 103rd Congress 1993: 67). He told the Senate Subcommittee on Defense, Committee on Appropriations,

Figure 11. Hawai'i's strategic location. From Hawaii Military Land Use Master Plan.

that "our area of responsibility stretches over 100 million square miles, about 52 percent of the Earth's surface in an area that encompasses two-thirds of the Earth's population, an area where we have relations with more than 40 nations in a coordinated strategy" (67). The military's mapping of Hawai'i in the world, like all maps, is both "instrument and signifier of spatialization" (Haraway 1997: 135). Taken at face value, the map is simply a two-dimensional representation of half the globe; read metaphorically, it is a naturalization/justification of the presence of CINCPAC in Hawai'i and of the United States in the world. "Maps are models of worlds," Donna Haraway reminds us, "crafted through and for specific practices of intervening and particular ways of life" (135). This particular mapping puts Hawai'i right in the middle of the action, suggesting that it is the logical place from which to control the surrounding, undifferentiated real estate. Admiral Larson employs the tools of administrative rationality and management by objectives to answer the question, "Will the United States remain a political, economic and military superpower?" (69). Assuming, of course, that "the answer to that must be 'Yes'" (69), the admiral reaches into his "tool kit" to produce the mechanisms "for engagement, for deterrence, for influence, and for rapid reaction" (79).

Through the lens of national security, in short, the world is a dangerous place. Admiral Larson performs in elite circles the same discursive maneuvers as does this nine-year-old child living in military housing, who spoke to a local reporter about the fenced compound with guarded access and homogenous residents:

> I love this housing because it is very safe. We don't need to worry about thieves, kidnappers or murderers. Before anyone enters the Peninsula gate, guards check to make sure whoever enters is a resident or visitor. Also, the guards patrol the Peninsula to make sure everyone's OK. (Quoted in *Honolulu Star-Bulletin* 1991: A-17)

The constraints on difference that enable this safety, and the price that such communities pay to enforce it, disappear into the peril marked by thieves, kidnappers, and murderers. This child's reflections are amplified in the national anxiety attack accompanying the Iran-Contra hearings, presided over by Senator Daniel Inouye. Cynthia Enloe points out that during the hearings, a persistent theme from all sides was "we live in a dangerous world":

> No one chimed in with "Well, I don't know; it doesn't feel so dangerous to me." No one questioned this portrayal of the world as permeated by risk and violence. No one even attempted to redefine "danger" by suggesting that the world may indeed be dangerous, but especially so for those people who are losing access to land or being subjected to unsafe contraceptives. Instead, the vision that informed these male officials' foreign-policy choices was of a world in which two super-powers were eyeball-to-eyeball, where small risks were justified in the name of staving off bigger risks—the risk of Soviet expansion, the risk of nuclear war. It was a world in which taking risks was proof of one's manliness and therefore of one's qualification to govern. (Enloe 1990: 12)

Relatively little recalibration has been necessary to carry this language of danger into the post–Cold War era, which Senator Strom Thurmond has called the "Age of Chaos" (1995: 34). As a 1991 editorial in the *Honolulu Advertiser* proclaimed, "Uncertainty seems to have replaced communism as the biggest threat to U.S. security interests in the Asia Pacific region" (Griffin 1991: F-3). Quoting Defense Appropriations Subcommittee Chairman Daniel Inouye that uncertainty is "our great challenge" and "the underlying issue," the editorialist, John Griffin, went on to set aside suspicions about military posturing in the face of feared budget cuts to give full support to Pacific Commander Larson's suggestion that

as superpower confrontations diminish in the 1990's, economic competition for resources and markets may heighten latent regional friction. In this context, the traditional role of the U.S. as the region's
honest broker assumes even greater importance. (Griffin 1991: F-3)

Even more interesting than the assumption of a disinterested United
States brokering power "honestly" is the portrait of uncertainty and
unpredictability as the greatest threats to national security. National
security discourse is both domestic and international in its scope and
methods. It creates and promotes a kind of public space in which
safety requires "the violent domestication of forces that are presented as external, alien, and in need of taming." Its effect is "to
inscribe a political space with violence" (Klein 1989: 99). Both the
creative possibilities latent in uncertainty and the ceaseless violence
required for the mastery of events disappear from view as the common currency for producing coherence is monopolized by the national security state.

Unlike war talk, which finds it difficult to disavow its debt to
peace, national security discourse has no attractive opposite to which
its debts to otherness might come due. While a man of war might
long for peace, what other state of affairs could a functionary of the
national security state be for? The answer, it seems, is *more* national
security. National security discourse goes on forever because, without radically altering the terms of the debate, it is difficult to define
an attractive alternative.[3] National security is simply the rational
path to the stability that, as Admiral Larson sees it, assures "continued access to the money, to the markets, to the resources, to the
trading partners in Asia" (United States 103rd Congress 1993: 71).
National security also allegedly promotes nonviolence: the admiral
assures the Senate subcommittee that "we do not want anyone to be
tempted to resort to force. We want that to be unthinkable" (75).

National security discourse also goes on forever because, dependent on its self-constitution vis-à-vis "danger," it is obsessed with establishing and protecting boundaries that can never be finally fixed.
The national security discourse provides a mode of ordering the unruliness "beyond the borders of the sovereign state; 'over there' in
bordering or far distant regions where strange forces of otherness
well up to challenge domestic order" (Klein 1989: 101). "Clearly the
world is one in which there is a continuous need to employ military

forces," declares strategic analyst Michael Vlahos (1995: 33). Taking this need as a self-evident starting point rather than an ontological commitment, strategic analysts anchor themselves in "facts" that they then labor ceaselessly to assure. In the national security view, the world is "a 'tough' world of power politics . . . presented as a fully articulated given" with no sense that such language assists in the creation of "that to which it purports to be responding" (Klein 1989: 100). National security discourse's pedagogical moment is the alleged givenness of this danger; its constant performative labor lies in the production and disguise of this self-evidence within the hegemonic "common sense" of a society. The same institutions that produce Admiral Larson's hope that peace will come by way of strengthened national security efforts also help to produce threats to peace through the massive and indiscriminate international trade in armaments (Sennott 1996: B-1–B-12).

War talk and national security discourse carry two distinct erotic charges. War talk enunciates a young and reckless masculinity, a martial brotherhood, an immediate phallic conquest of risk, activation of loyalty, quest for victory. National security discourse's seductions are somewhat more sanitized, calling on a "drive to superior power" (Cohn 1987: 697) that combines the impersonalities of risk management with the erotic enthusiasms of sleek, high-tech weaponry. War talk recruits us into the soldier's world, where young men, and sometimes women, can become impassioned about duty, heroism, or just plain "kicking butt," while citizens readily rally to "support our boys." National security discourse invites us into a more sophisticated arena, where somber middle-aged men calculate strategic political and economic advantages through an erotics of domination and control. If war talk evokes the image of a college fraternity run amok, national security discourse is more akin to a well-ordered system of organized crime.

In the implicit tension between the discourses of national security and of war, the predominance of the first is not a natural phenomenon but a politically aware choice of self-representation by the military. That President Eisenhower's warnings about the dangers of the military industrial complex appear both accurate and quaint is testimony to the distance the United States has traveled toward the national security state since World War II.[4] Yet the division between the two discourses is not so much a wall as a membrane, permitting leakage from one side to the other while still enforcing a distinction (Connolly 1995: 145). One aspect of the leakage is the shared internal doubleness of the two ways of framing militarization. Returning to Doty's understanding of double-writing, both national security discourse and war talk simultaneously assert what is real and disguise the rhetorical labor that such assertions require; that is, they both "move ambivalently between the pedagogical and the performative" (1996a: 179). Claude Lefort maps the political power needed to maintain the necessary and problematic relation between givenness and invention:

> This exorbitant power must, in fact, be shown, and at the same time it must owe nothing to the movement which makes it appear. To be true to its image, the rule must be abstracted from any question concerning its origin; thus it goes beyond the operations that it controls. Only the authority of the master allows the contradiction to be concealed, but he is himself an object of representation; presented as possessor of the knowledge of rule, he allows the contradiction to appear through himself. (Quoted in Bhabha 1990: 298)

The pedagogical moment of both war talk and national security discourse entails what Doty calls "the territorial trap": the assumptions that "states are fixed units of sovereign space," that "the domestic/international dichotomy is unproblematic," and that "states are spacial containers of societies" (1996a: 172). Beyond these givens, national security discourse elaborates the extended perception of danger that both fuels and undermines its quest for order. The ceaseless performative labor of the national security state lies in the production and disguise of the obviousness that surrounds its pedagogical assumptions. The assertions by Admiral Larson and countless others that our common reality is one of uncertainty and danger teach citizens, by example and repetition, The Big Is. The charts, graphs, calculations, plans, the sophisticated lingo of "forward presence" and "crisis response" are the performative techniques that make possible the assumptions that The Big Is requires. At the same time, the double-writing of the pedagogical/performative "threatens to undo these very assumptions" through practices that reveal the fluidity of that which was taken as given (Doty 1996a: 179). The continuous sedimentation of ordinary so-called common sense asserts the uncontestability of danger, its foundational presence;

repetitious performative strategies do the necessary work to secure the obviousness of danger, while ironically calling attention, through their busy labor, to the need for such securing in the first place (see Bhabha 1990: 297). In the gap between "the iconic image of authority and the movement of the signifier that produces the image," there is a "space of liminality" where the work necessary to assure the appearance of certainty also signifies certainty's potential collapse (Bhabha 1990: 300).

An example of the doubled doubleness of war talk and national security discourse, of the pedagogical and the performative, looms in the discourse and institutions of international arms sales. As the Pentagon has bought relatively fewer arms in the post–Cold War era, U.S. arms manufacturers, eschewing conversion to less bellicose production, have turned abroad, making the United States "the leading exporter of conventional weapons" (Sennott 1996: B-2). Government-to-government sales by the United States account for 64 percent of the global market; this number does not include "direct commercial sales" between U.S. manufacturers and foreign states, which may total another $6 billion (Department of Defense report, quoted in Sennott 1996: B-2). International arms bazaars such as the Dubai Air Show, "the Third World's largest showcase of the latest high-tech weaponry," offer sites for the merger of war talk and national security discourse into an effective sales pitch (Sennott 1996: B-2). Jaunty American pilots sporting Gulf War decorations team up with marketing experts from the arms industries to impress potential customers who have the millions/billions of dollars available to purchase the latest in fighter jets, attack helicopters, missiles, and tanks. The presence of dashing pilots in full military regalia, boasting about the weapons' successful performances against Iraq, brings a dash of war talk into the stew of national security discourse. "Our role here is national security," says Air Force Major Todd Freuhling of the international sales, emphasizing the alleged "regional stability" such sales are meant to secure. But some of the exhibitionist pleasures of the weapons traffic also peek out: "Quite frankly, we are putting our hardware on display because it's the best in the world and we're proud of it" (quoted in Sennott 1996: B-2).

Investigating the complex relations between pedagogical and performative allows one to see the discursive and institutional circumstances of production of the given. The international arms trade connects the pedagogical (the litany of "it's a dangerous world") with the discursive and material construction of the circumstances of such danger (the United States and other nations are making the world more dangerous by selling all these weapons). It is not helpful to intervene in this dance of pedagogical and performative by simply refusing the pedagogical claims; it does not get one very far, vis-à-vis claims about danger, to simply deny that there is danger. Indeed, it is a dangerous world, and the United States assures it will be so by exporting billions of dollars in weapons to those very regions whose potential instability, and accompanying inclination to violent suppression of internal critics, fuels Pentagon commitments to prepare to fight in two simultaneous regional conflicts. Suggestively named "blow-back" in military talk, U.S. troops have faced American-made weapons "in virtually every conflict into which the US has sent troops since the 1989 collapse of the Soviet Union" (Sennott 1996: B-6). During the 1980s, the United States supplied $200 million in weapons to the government of Somalia, then spent $2 billion and saw twenty-four American soldiers die in subsequent "peace-keeping" efforts (Spain 1995: 47). "We end up fueling the conflicts we seek to contain," comments director for the Institute of Defense and Disarmament Studies Randall Forsberg (quoted in Sennott 1996: B-3).[5]

The arms trade also produces insecurity in the form of loss of jobs for American workers. Despite promises otherwise, jobs are often transferred overseas along with the weapons. Known in the trade as "off-sets," such deals transfer production jobs along with weapons systems to the purchasing nations (Sennott 1996: B-11). Pentagon estimates place the loss of U.S. defense jobs at one million since 1989 and predict the pattern will continue (Sennott 1996: B-11). A local union official in Lynn, Massachusetts, where the General Electric plant sent jobs to the Netherlands as part of a sale of helicopter engines to the Dutch monarchy, commented bitterly, "They stopped doing bribes in cash, now they do it in American jobs" (quoted in Sennott 1996: B-11). Meanwhile, "leading defense manufacturers report big profits, surging stock prices and huge increases in compensation for top executives (an average of $3.5 million a year for the 10 biggest companies)" (Sennott 1996: B-2). While "the Pentagon is a very costly employment agency," the combination of shrinking jobs and expanding profits suggests a capital-friendly, labor-unfriendly security strategy (Moore 1995: 37).

Enhanced profits for arms manufacturers translate into giant leaps in the insecurity of the most vulnerable Americans. Public monies for schools, hospitals, and other social services shrink while tax dollars subsidize the arms trade. Weapons profits are enhanced by the U.S. policy, begun under President Bush, to waive the "re-coupment fees" originally intended to reimburse public coffers for the massive amounts of tax dollars ($36 billion a year under the Clinton administration) spent on weapons research, development, and testing (Sennott 1996: B-2). Other direct subsidies include the Pentagon's annual $3.5 billion "foreign military finance" program, which gives money to "friendly" nations for the purchase of arms. Tax dollars also go to support industry mergers such as those between Lockheed and Martin Marietta, between Boeing and McDonnell Douglas, and between Raytheon and Hughes Aircraft: public monies subsidize the CEO bonuses, the companies' consolidation costs, and the subsequent loss of jobs to U.S. workers (Vartabedian 1997: B-8; Hartung 1997: 6).[6] Public monies provide loan guarantees for arms purchases and marketing presentations at shows such as Dubai: "The United States spends more than $450 million and employs nearly 6,500 full-time people to promote and service foreign arms sales by U.S. companies" (Ivins 1996: A-8). Money that could have been used to build schools or hospitals disappears into the giant maw of arms profits. Nobel Peace Laureate Oscar Arias Sánchez noted:

> It is the greed of the arms trade that threatens our common future. In a world where 900 million adults do not know how to read or write, 1 billion people do not have access to potable water, the arms merchants bear much of the blame for this poverty. (Quoted in Sennott 1996: B-3)[7]

At the same time, the arms trade produces a certain kind of security for the arms producers themselves; it virtually guarantees the demand for more and better weapons in the near future to contest those now for sale. Lawrence Korb, former Assistant Secretary of Defense under Ronald Reagan, now a senior fellow at the Brookings Institute, says, "it has become a money game: an absurd spiral in which we export arms only to have to develop more sophisticated ones to counter those spread out all over the world" (quoted in Sennott 1996: B-2). While Korb's concern that this trend "undermines our moral authority in the New World Order" might be somewhat belated, he correctly points out that U.S. officials, under these circumstances, will have little success convincing other governments to sell fewer arms (Sennott 1996: B-2). In defense of the expanded arms trade, a spokesman for the weapons manufacturers' chief lobby, the Aerospace Industry Association, claims: "If you look at other markets, like medical instruments, supercomputers, et cetera, the US has 70 to 80 percent. The question isn't why do we have so many exports in the aerospace market; the question is why have we had so little?" (Sennott 1996: B-2). With the Pentagon's blessing, weapons manufacturers merge and remerge in order "to make U.S. firms more competitive in the global marketplace" (Diamond 1997: B-7). The normalization of weapons sales as just another business deal, like the normalization of the military as just another job, helps to hide in plain sight the material production of the circumstances that produce the pedagogy of the national security discourse.

The international word traffic in "terrorism" offers another example of the production of danger that appears to be simply found in ways that make it seem self-evident rather than socially produced. Terrorism is a kind of unlawful violence: violence by those who are not (yet) states, violence that does not play by the rules. Terrorism is outside the boundaries of the legitimate use of force as states have established them. The pedagogical given here is, as Doty remarks, that states are sovereign spatial containers of legitimate societies, with clear boundaries of inside and outside, with guns and the right to use them. Acts of war against states by nonstates are not war but terrorism, which denies the legitimacy or potency of the boundaries that states are constantly reinforcing. Yet the terrorists are emulating the means by which states initially became states, the messy beginnings that are quickly abstracted from political argument once mythic origins are secured. And the goal of the terrorists is often to follow the lead of the states, to become states themselves. The state system produces the terror by monopolizing legitimacy and excluding from the picture both its nonstate others and its own story of how it became a state. The performative move here includes all the effort that goes into making and hiding these exclusions, naturalizing the boundaries that are nonetheless ceaselessly negotiated and renegotiated.[8]

Such effort becomes temporarily visible in circumstances such as the disclosure of CIA drug deals with the Nicaraguan Contras, thought of as terrorists by those loyal to the Sandinista government. Periodic public allegations of CIA involvement in cocaine traffic into the United States in order to secure funding for illegal weapons shipments to the Contras might temporarily reveal the performative at work behind the pedagogical, thus questioning the allegedly stable boundaries between inside and outside, between legal and illegal, between policy and crime. Ultimately, the legitimacy of the state itself might unravel. The CIA, supposedly the brains of the national security apparatus, appears as less than helpful in monitoring danger and guarding our nation by collecting needed intelligence. But it excels at cutting deals with drug cartels, suppressing popular protest movements, and illegally channeling weapons into other nation's civil conflicts.[9] The naturalization of danger becomes unanchored when one can see the performative at work: the arms merchants' production of social, occupational, and military insecurity, which then legitimizes further arms production; the state system's production of terrorism, which then legitimizes extensive surveillance and regulation of citizens; the CIA's contribution to producing the drug traffic, which then legitimizes the massively expensive and destructive "war on drugs." In these moments of contradiction between the pedagogical and the performative, the dangerous politics of national security discourse become more visible and more vulnerable to critique.

National security discourse operates as a kind of artful dodger behind its more visible and in some ways more vulnerable partner, war talk. While both war talk and national security discourse are produced out of the interaction between what can be taken for granted and what is necessary to construct the taken-for-granted, war talk operates more fully within the pedagogical and is thus more vulnerable to having its productions flushed out of hiding. National security discourse is tilted more toward the performative, thus better at hiding its productions. Our goal is to tease out the intertwining doubleness of national security discourse and war talk, and of the performative and the pedagogical, to leave them no place to hide, in order to challenge the military's near-monopoly on the meaning of citizenship in contemporary American politics. The practices involved in a more democratic citizenship require toleration of some of the central demons in the national security universe: instability, fluidity, difference, openness to multiple possibilities. Democratic contestation is messy. It requires holding political arrangements open, subjecting them to questioning and recalibration; it entails some acceptance of an unpredictable world.

EXCESS AND LACK (NOT) AT FORT DERUSSY

The intertwined voices of war talk and national security discourse combine to do their work in the Fort DeRussy narrative. The glories and the sacrifices hailed in war talk appear prominently in the first gallery's presentation of victory celebrations at the end of World War II, in scattered photos of local citizens buying savings bonds or tending Victory Gardens, in photos of local children playing soldiers or ethnic groups bonding together in a common cause, in the reconstruction of a bar scene in Vietnam and of the daily lives of soldiers, and in the solemn final Gallery of Heroes. National security discourse comes more visibly onto the scene in the close scrutiny of military techniques, the evaluation of strategy and tactics, and most of all in the historical narrative of Hawai'i's relation to the western world. The discourse of national security often appears to be seamless, completely self-referential; it disguises its debt toward otherness by bounding its discursive space with mirrors. In this it is much like Edward Said's description of Orientalism, which

> shares with magic and with mythology the self-containing, self-reinforcing character of a closed system, in which objects are what they are *because* they are what they are, for once, for all time, for ontological reasons that no empirical material can either dislodge or alter. (1979: 70)

The self-referentiality of national security discourse continues that of Cook's narratives; remember that Cook and his sponsors had determined what land masses and water passages ought to be findable prior to any looking and then took their own story as evidence for its validity. Our interviews with public affairs officers on several O'ahu military bases were laced with self-referential understandings; questions about the military's relation to Hawai'i were answered with observations about the military's relation to itself. Praise of heterogeneity, such as that encountered when young soldiers travel to distant lands and meet unusual people, was framed within a very homogeneous context. One public affairs officer noted with some pride that "when we come into a place, we kind of adopt it" (interview with Pfister 1991). Another noted that the military gets to know a place, settles in, and adjusts it to themselves (interview with McCouch 1991). One might call this an example of what William James called viciously acquired naïveté: military representatives call it knowing the place, when it looks more like a brief brush with a place as it is forced to know them.

The production of this self-referential discourse is hidden in plain sight in the interactions between the pedagogical and the performative moments in the production of truth claims. The tutorial givens that animate the museum's narrative include the implicit assumptions in war talk about masculinity and soldiering that make the dead into heroes, rather than, say, victims or killers or dupes or confused young men with mixed motives. The pedagogical moment further occurs in national security discourse's historical telos, which takes as given that Hawai'i is part of the United States, to be understood in relation to the United States. The performative strategies in the museum include all those rhetorical gestures and grammatical devices that produce the Hawai'i that the military needs. Other aspects of Hawai'i, its competing histories and politics, constitute a noneligible otherness, "a situation of excess, in which the meanings and identities crucial to [the dominant] assumptions cannot be contained or sustained" (Doty 1996a: 176). Some excess is inevitable in any narrative because the relation between the pedagogical and the performative can get unruly: making and hiding what seems simply to *be there* always contains a glimpse of the "there" as made and the making as hidden. Hawai'i's excesses are, metaphorically, gendered feminine in relation to the hegemonic discourses of militarism in much the same way that Hawai'i's geographic and cultural differences were gendered feminine by the early discoverers. "Too much" and "too little" (excess and lack) are, after all, the flip sides of one another: neither fits; both disrupt; each can invite conquest or conversion by the combined anxieties and opportunities it ignites.

Naming examples of this excess highlights the dominant narrative's efforts at containment by attending to its loose ends, its valiant efforts, its stubborn moments of incompleteness. Such messiness flags sites of ongoing deferral, places where the military's story of itself in Hawai'i becomes more obviously intertwined with contending voices and resistant material. These sites of deferral are themselves, of course, unstable discursive formations, hosting their own debts to otherness, their own necessary deferrals, their own dilemmas concerning boundaries and exclusions. Our point is not to find a stable or innocent reality behind militaristic appearances, or a radical "'transcendental signified' where the process of deferral somehow would come to an end" (Moi 1985: 106). Rather, by pointing to the necessary incompleteness and complicated interdependencies of all claims to meaning, we are arguing for a politics that attends to heteroglossia and honors difference.

Some aspects of the implicit unruliness of Hawai'i's spaces could be its unique geographic bodies of land and water. While active volcanoes, annual hurricane seasons, and the occasional tidal wave are not unique to Hawai'i, the potent combination markedly differentiates Hawai'i from other parts of the United States. It is difficult to ignore the materiality of land, sky, and ocean, conventionally coded as feminine Nature vis-à-vis masculine Culture. Palm trees and beaches, endlessly represented as seductions in the international discursive economies of tourism, are sites for more than sunbathing and snorkeling: they are powerful ongoing scenes for local resistance to "development." The body of law in Hawai'i adds certain indigenous principles of land use to the western property laws that otherwise dominate. All beaches, except those controlled by the military, are public up to the debris line, the highest reach of the waves during the highest tide season (Hawaii Revised Statutes 1993: Ch. 205 A-1). The Hawai'i Supreme Court has suggested in *Public Access Shoreline Hawaii (PASH) v. Hawaii County Planning Commission* "that every development permit issued to a property owner requires determination of the existence of native-use rights" (Boylan 1996: A-10). In addition to sending "developers" into fits of anxiety, these and other rulings have stimulated a number of grassroots efforts to preserve beaches and their surrounding locales from construction and privatization.[10]

Like geographic bodies, human bodies have a somewhat different public presence in Hawai'i. Men and women in suits look a bit out of place in many social spaces, where "aloha wear" and mu'umu'us both commercialize a local custom and transgress a national one. Less clothing reveals more skin in a rich range of colors. The most popular and successful athletic event is wahine (women's) volleyball, which plays to sold-out audiences at the university stadium despite the game's availability on local cable TV stations (Chadwin 1996).

Other popular sports include women's softball and women's basket-
ball; a visible presence of strong and active female bodies is a regular
part of public culture. Surfing, boogie boarding, kayaking, canoe
racing, skin diving, and snorkeling take female and male bodies into
public exposure in ways relatively uncommon on most of the mainland, connoting both unusual sports and active border crossings between land and water.

Other sorts of public bodies manifest themselves in the local sex
trade, much of which addresses itself to a military clientele. In a recent study of prostitution on Oʻahu, RaeDeen M. Keahiolalo (1995)
found that nine out of the ten prostitutes she interviewed said that
military men made up more than 60 percent of their customers. The
six military men she interviewed estimated that as many as 90 percent of military men participate in prostitution while on deployment.
The five social workers she interviewed, most of whom work with
sexual laborers, agreed that prostitution is more extensive in areas
occupied by the military. While this study provides only an impression, it nonetheless gestures toward an explosive issue that the military and its civic boosters in Hawaiʻi are understandably reluctant to
explore. When the huge U.S. military presence in Okinawa became
controversial, catalyzed by the highly publicized rape of a young girl
by three servicemen, Governor Benjamin Cayetano was quick to welcome the transfer of those troops to Hawaiʻi.[11] Evidently unmoved
by the possibilities of inviting similar actions in his home state, the
governor also failed to interpret the warning embedded in Admiral
Richard Macke's much publicized remark: "For the price they paid
to rent the car they could have had a girl" (quoted in Landers 1995:
A3). Amid the uproar occasioned by the CINCPAC commander's
casual recognition of prostitution around military bases, few have
noted that the former commander was right—the soldiers could have
easily and cheaply "had a girl." Governor Cayetano's eagerness to
host more U.S. troops in Hawaiʻi overlooks the likely expansion of
the sex trade that already flourishes in parts of Chinatown, Waikīkī,
Wahiawā, and Hotel Street.[12]

Still other kinds of bodies have become visible in the highly publicized trial of *Baehr v. Lewin,* in which three gay couples sued the state
for the right to marry.[13] While public opinion polls find that a majority of residents of Hawaiʻi oppose same-sex marriage, a significant
minority supports it. In hosting this controversial challenge against

compulsory heterosexuality, the state is continuing its relatively liberal civil rights traditions, including its early legalization of abortion in 1970 and its prompt ratification of the Equal Rights Amendment "less than two hours after the Congressional vote" (O'Donnell 1996b: 14). In its initial ruling in *Baehr v. Lewin* the First Circuit Court of Hawai'i interpreted the case within Hawai'i's "history of tolerance for all peoples and their cultures" (quoted in O'Donnell 1996b: 14).

The colonial traffic in tropical bodies that produced Hawai'i's multiethnic present also marks another arena of excess in relation to the discourses of militarism. Certainly most localities in the United States could narrate a particular, layered history that defies the flattening of meaning imposed by militarized narratives (see, e.g., Michael J. Shapiro 1997; Pratt 1984). Yet Hawai'i is unique in that its colonizations are relatively recent, and its racial geographies quite diverse. The constant production and reproduction of what it means to be "local" are an indication of both anxiety about and resistance to hegemonic structures of meaning. The *Honolulu Advertiser* ran a series in its Island Life section asking readers to contribute their definitions of *local,* which many did; a regular column in that same paper featured "things local" and was taken to task for offending tourists; the publication *Bamboo Ridge* hosts animated ongoing debates about what does and does not constitute local literature. The pros and cons of speaking pidgin (a rich and musical stew of English, Hawaiian, and various Asian languages) are hotly contested. Local food, local footware, local music, local drama, local poetry, local fashion—each of these is a publicly available site of excess. They do not fit. In these ways, Hawai'i itself is a noneligible otherness facing the national security state.

Racial and ethnic turbulence within the arena of the local complicates the discursive terrain without undoing the friction vis-à-vis military spaces. *Local* can be thought of as meaning "from this place"; descendants of the original Polynesian settlers and of the imported workers from Japan, China, Korea, the Philippines, Portugal, and elsewhere are, while distinct in many ways, local.[14] *Local* has often come to mean, in light of Hawai'i's history, not white. Yet there are, of course, many haoles (the Hawaiian word for outsider that is used for most Caucasians) who are "from this place" in the sense that they were born in Hawai'i; the term *local haole* is sometimes used both to designate a white person who is from this place, and, simultaneously, to retain the primary nonwhiteness of localness.

Most military personnel and their families in Hawai'i are haoles from the mainland, while a very visible military minority is African American. To be haole in Hawai'i is to be marked, racially, as part of a group whose history is one of conquest and conversion. Most white people from the United States are unaccustomed to thinking of themselves in racial terms, since their political and economic privileges have operated to reassure them of their universal status; white, in other words, operates in most mainland contexts as the norm, the category that requires no explanation (see Pratt 1984; Frankenberg 1993; Whitaker 1986; Morrison 1990). The historical erasures that have allowed most mainland whites to think of themselves simply as "Americans," rather than as descendants of a group who eliminated a prior claim through violence, are less fully institutionalized in Hawai'i. The raw and recent histories of colonialism, along with its continued reproductions and energetic contestations, make it difficult for white people in Hawai'i to avoid being marked as white. *Haole* can be a matter-of-fact descriptor or (depending on what adjectives precede it) a bitter denunciation, but it seldom disappears into an implicit universal. Unaccustomed to being marked, racially, as white, Caucasian service members may view the category "haole" simply as derogatory, but usage is far more complex (see Kubo 1997; Rohrer 1997; Whittaker 1986). Colonial politicizations of whiteness in Hawai'i enable the vigorous category of "local" to contest the racial implications of a largely Caucasian military presence.

Hawai'i's local excess does not translate into an obvious unified resistance to militarized order. On the contrary, local people interact with the military presence in ways that are much more complicated than straightforward rejection, or happy acceptance.[15] The local National Guard is an institutional site that many local men and women turn to their advantage, acquiring therein a needed second or third job and a tuition waiver for the University of Hawai'i.[16] More than four thousand civilian workers are employed at Pearl Harbor shipyards, and while some of them are brought in from the mainland, most are local skilled and unskilled laborers.[17] Several privately owned shipyards rely on federal contracts, rendering more local workers dependent on military spending. Some local small businesses depend on military customers. Our point is not that local people and local cultures are unanimous in their opposition to the military in Hawai'i but, rather, that the rich multiple sites of localness are themselves excessive vis-à-vis a militarized order.

By ordering service members to Hawai'i, the military significantly changes the demographic arrangements of the state. When the Fifth Cavalry Regiment occupied Schofield Barracks in 1909, "it was the greatest influx of American blood" Hawai'i had ever witnessed (Joesting 1972: 282). Consequently, because military personnel are a strong minority in Hawai'i, comprising with their dependents approximately 16 percent of O'ahu's population, they have the ability to change the unique character of the islands and to force the local culture to become more "acceptable" (Lind 1984/85: 25). At the same time, service members are often segregated from the general populace because an "enormous military ghetto is created with complete on-base housing, shopping, and recreational facilities" (Gray 1972: 15). The average tour of duty in Hawai'i is three years, but much of this time is spent at sea or on deployment to East Asia. This leaves little time or initiative for service members to acculturate themselves to the local scene. Military personnel are often indistinguishable, except perhaps by their distinctive haircuts, from the tourists.

The borders separating military personnel from the surrounding community are of course permeable, hosting multiple kinds of commodity and labor exchanges across militarized boundaries. Yet the military's self-understanding is firmly located within what Michael J. Shapiro calls "the sovereignty impulse," which "tends toward drawing firm boundaries around the self, in order unambiguously to specify individual and collective identities, to privilege and rationalize aspects of a homogeneous subjectivity that is eligible for membership and recognition." Sovereignty practices, by this understanding, work "to constitute forms of nonidentical and ineligible otherness," and to name and contain "both the spaces in which subjects achieve eligibility and the space in which the collective as a whole has dominion" (Shapiro 1993: 2–3). Faced with Hawai'i's otherness, the military plays the sovereignty card with a vengeance: it draws firm boundaries around itself, specifies clearly who is in and who is out, privileges and naturalizes its own self-understandings, and constitutes Native Hawaiians and other local people as either useful and recruitable, or as outside the boundaries of eligibility. Interactions with locals are orchestrated on a public relations stage that is controlled much more by military interests than local ones. For instance, troops stationed in Hawai'i are sometimes utilized for "good neighbor" activities such as repairing schools or assisting with hurricane relief, but such crossings into local territories are done on the military's terms. It is the military that controls the resources that then may (or may not) be used to provide local assistance. If the schools were adequately funded in the first place (rather than losing out to the military in competition for scarce federal funds), they would not need military largesse. Of course, residents are grateful when the military steps in to meet a pressing community need, as did the Seabees in July 1997 when they replaced a collapsed bridge in Moloa'a Valley in Kaua'i. While state and county officials squabbled over whose bridge it actually was, the Seabees replaced the bridge in three days for less than $10,000. Amid the enthusiastic appreciation expressed in front-page coverage of the event, few questions were raised about the distribution of resources enabling the military's generosity (Ten-Bruggencate 1997: A-1).

Service members' perception of the state's residents seems to typify a perception of otherness generally associated with foreign nationals, an attitude that "we don't have anything in common with them and they don't have anything in common with us" (Menton and Tamura 1989: 319). Two high-ranking military officers commented to us that Hawai'i is widely considered to be "overseas duty" (interview with Keefer 1991) or "a foreign assignment" (interview with Imai 1995). At least in some cases relations between civilians and military personnel are conditioned by "the attitudes of racial superiority freely expressed by the men in the service" (Lind 1968: 252). Interactions between residents and military personnel sometimes produce overt violence between the two communities. The Damon Tract Riot of 1945 (nowhere mentioned in the Fort DeRussy Army Museum) pitted approximately one thousand sailors from the Honolulu Naval Air Station against nearby civilians, termed "gooks" by the sailors (Lind 1968: 258).[18] Since Damon Tract, smaller conflicts have continued, resulting in a furthering of ill will on the part of both groups. A 1991 New Year's Day column by *Honolulu Star-Bulletin* contributing editor A. A. Smyser refers to "fights between 'locals' and servicemen" as among the problems in his early experience of Hawai'i (Smyser 1991: A-10). Occasional acts of violence by local people against service members, like those against tourists, are highly publicized and often framed in political terms, as ominous local assaults on harmless and defenseless outsiders.[19] Such attacks engender considerable public hand-wringing by officials, while violence by service members is treated as individual misbehavior rather than as a consequence of an institutional presence.

More pervasive than the occasional physical confrontation between military personnel and local people is the general policy of separateness and the casual disregard for local culture and history combined with an equally casual willingness to appropriate their symbols and resources. Military authorities selectively recruit certain local histories and indigenous symbols as accessory possessions for military use, domesticating potential excess by making it fit into the hegemonic order. The insignia of the 25th Infantry Division (Light), based at Schofield Barracks, for example, is the taro leaf with a lightning bolt emblazoned over it in bright red and yellow (the traditional colors of the Hawaiian monarchy). The army hotel at Fort DeRussy, the services' main recreational center in the state, offers packaged, popularized Hawaiian culture à la Waikīkī and is fitted with a Hawaiian name—Hale Koa (literally, the house of the warrior). Navy ships arriving at Pearl Harbor are greeted with hula dancers on dock while gigantic plastic leis are hung over the ship's bows; Hawaiian words of greeting (*aloha*) and thanks (*mahalo*) dot the speeches of incoming commanders (interview with Harris 1991).

Military spokespersons also make Hawai'i appropriate for themselves by inserting their history into the history of local immigration, representing the military as simply one more unremarkable arrival. One military public affairs officer remarked to us, "Nobody's from here. So everyone's an outsider in Hawaii" (interview with McCouch 1991). This configuration of border crossings eliminates critical differences between the Polynesians who voyaged here two thousand years ago, the missionaries and plantation laborers who arrived in the nineteenth century, and the soldiers who just got off the boat. Claims that everyone in Hawai'i is an immigrant serve to normalize the military's largely Caucasian presence and to erase the particular histories of labor immigration, conversion, and conquest.

Comparable erasures and lingering excesses inhabit the periodic struggles over local naming practices. Because "many of the Marines have had difficulty pronouncing the names of the Hawaiian Islands," the Marine Barracks Hawaii, located at Pearl Harbor, held a "Rename the Islands Contest" in 1990 (Marine Barracks of Hawaii Monthly Newsletter 1990). Some of the winning entries included "French Island" for the island of Hawai'i (named in honor of the commanding officer of the barracks) and Bush Island for Kauai'i (named after President George Bush).[20] These marines were following in the footsteps of their military fathers in claiming the power to discipline Hawai'i's linguistic excess. McClintock comments that "the desire to name expresses a desire for a single origin alongside a desire to control the issue of that origin" (McClintock 1995: 28). Because the military initiated much of the official surveying and mapping of Hawai'i, they "haolefied" the names or perpetuated existent haole names.[21] In Pearl Harbor (Pu'uloa) numerous areas now have English designations: Bishop Point (Lae o Halekahi), Hospital Point (Lae o Keanapua'a), Ford Island (Moku'ume'ume). The area previously called Kahauiki has disappeared altogether with its incorporation into Fort Shafter Military Reservation. In 1948 the army changed the name of the Mokule'ia Military Reservation to Dillingham Military Reservation, disparaging another Hawaiian word "that's too hard to pronounce" (Honolulu Star-Bulletin 1948: 5). These acts of casual appropriation do more than insult the local people, although they accomplish that outcome also; they knit space into discourse differently, enabling new sets of linguistic relations and disabling old ones. They transform Hawaiian spaces into objects of military knowledge. The military's renaming of Hawai'i operates "as the agent of a linguistic fifth column, infiltrating and dividing the space stealthily, as an outpost supplying a ramifying network of grammatical and syntactical connections" (Carter 1987: 58). Militarized names teach the landscape to speak differently, to offer itself for military use, to capitulate to possession, to cease reminding possessors of prior claims. Military renaming arrests Hawaiian spaces, enclosing them in a network of approved associations, neutralizing their otherness (Carter 1987: 61). Even the pressure of the territorial government, which, in 1929, asked the War Department to rename Fort Weaver to Fort Kalākaua, in honor of Hawai'i's last king, did not hold up against the military's "lordly power of bestowing names" (Nietzsche 1956a: 160).

Another space for the expression of Hawai'i's excess is the abundance of historical and archaeological sites, many on military land,

and the political struggles they ignite. Many Hawaiian activists link preservation of these sites to the maintenance of their culture and religion. The types of important sites over which the military exercises stewardship include burial sites, *heiau* (religious sites), *ko'a* (fishing shrines), *loko i'a* (fish ponds), *lua* (habitation and shelter caves), and *haka* (terraced platforms). Official military preservation of selected sites aims at collecting bits of scientific information for research purposes rather than preserving living sites of cultural meaning. The fishponds in Pearl Harbor were filled in, dredged, or left untended until they became "mudholes" (*Honolulu Advertiser* 1946: 5).[22] A 1979 study completed by the army inventoried a total of seventy-seven archaeological sites. These included the *Ukanipō Heiau* on the Mākua Military Reservation, the *Āliamanu* Burials at Aliamanu Military Reservation, and the *Hanaka'oe Haka* at Kahuku Training Area. Of the seventy-seven sites, forty were still in existence, five had been "relocated," and thirty-two had either been destroyed or could not be found (USASCH 1979: 102). Although the army admits that its actions contribute to the destruction of the sites (e.g., "The use of a large part of Mākua Valley as an impact area since 1943 has adversely affected three archeological sites"), it still believes that "Army protection reduces the likelihood of damage to the sites" (USASCH 1979: 138, vii). The pecularity of this notion of protection becomes less puzzling once one realizes that it is not cultural artifacts but information that is being protected: "Continued preservation for many sites may not be necessary; only the preservation of site information is essential. Detailed recording, written description, photographs, scaled maps and drawings, and possible limited salvage excavations could constitute adequate mitigation of any subsequent site destruction" (USASCH 1979: 139). The danger, for the navy, is not that assault on a way of life will be continued, but that scientific data might be lost. Native Hawaiian cultural and spiritual otherness is rendered undisturbing by folding it into a scientific model.

Perhaps the most highly politicized and public excess that evades and contests a militarized order in Hawai'i is represented by some of the Native Hawaiian political groups. The Native Hawaiian sovereignty movement, hosting complex contending perspectives on indigenous identities, national memberships, and control of land, provides a space in which the flow across boundaries denaturalizes the ontological status of the boundaries themselves. In contrast, the Fort

DeRussy Army Museum is a discursive site for what Doty calls "the ontological commitment to the state":

> The ontological commitment to the state ensures that the starting point is the existence of boundaries that are then transgressed, rather than the always-in-process practices that effect the construction of contingent, and never finally fixed, boundaries. (Doty 1996a: 176–77)

The sovereignty movement calls into question a number of boundaries and agitates a variety of familiar assumptions, raising questions about how to interpret ethnic differences, about who counts as a citizen, about how space is to be named and ordered. The internal politics of the movement are complex and dispersed, very much in process, offering a number of different interpretations of what sovereignty might mean and how it might be achieved.[23] The movement draws on a long history of resistance to the U.S. government, including numerous rebellions against the initial colonizers, against the dominance of foreigners in the monarchy, against the overthrow of Queen Liliʻuokalani, and against Hawaiʻi's incorporation into statehood.[24] None of these rebellions is mentioned in the Fort DeRussy narrative; nor is the contemporary challenge posed by the growing movement for some form of native sovereignty. The process of enforcing Fort DeRussy's silences offers a glimpse of "statecraft as the never finally completed project of working to fix meaning, authority, and control" (Doty 1996a: 177).

Sovereignty itself, no matter how defined, is not an unproblematic claim (see Shapiro 1991). In some ways, it is easily recuperated into the dominant narrative; it does, after all, seem to aspire ultimately to the establishment of new territorial and political boundaries, not to the exchange of meaning across boundaries. But Hawaiian notions of sovereignty are not yet fixed. They remain in contest, offering the possibility of "a fluid and deterritorialized understanding of space that is at least potentially incompatible with the static notion of space inherent in the modern system of territorially based nation-states" (Doty 1996a: 178). As a contemporary flow across existing boundaries, the sovereignty movement features discourse's performative moment by contesting the givenness of spatial arrangements. The voices in the sovereignty movement "denaturalize the space of the sovereign state by raising issues of political authority, political community, and national identity—things that governments would like to have taken for granted" (Doty 1996a: 178).

The movement operates in the ambiguous margin between pedagogical and performative, confounding the foundational essence that the Fort DeRussy narrative writes on Hawai'i, even if that narrative is sometimes inscribed in the name of a competing foundational essence.[25] Other political movements, some of them related to the sovereignty campaign, others anchored elsewhere, operate in this ambivalent space. A full-page ad in the *New York Times* on July 7, 1995, sponsored by the Hawai'i Ecumenical Coalition, asked, "Is it time to give back Hawai'i?" This ad was part of the Ecumenical Coalition's ongoing efforts to restore resources and dignity to Native Hawaiians (see Pennybacker 1996). The coalition was also active in protesting the Star Wars projects at Barking Sands missile-launching facility on the island of Kauai'i.[26] The American Friends Service Committee (AFSC) organizes periodic antimilitary protests, including an annual Caravan for Peace to "focus on the historical, economic, social, and environmental impacts of the massive presence of the US military and envision what a demilitarized future might be" (n.d.; see also Barrett 1996). The AFSC also produces and distributes a brief pamphlet called *Hawaiian Lands and the Military*, describing the historical processes by which the U.S. military has gained control of "ceded lands," which are crown and government lands of the Hawaiian monarchy that were taken by the federal government's annexation of Hawai'i. These lands are often sites of contest among the state, the military, and "developers." The political battles fought on ceded lands both draw on and stimulate the sovereignty movement, constantly irritating the seamlessness of the DeRussy narrative with contending interpretations of who is entitled to speak and what needs to be said.

Some of these land battles are relatively well known to the general public, due to persistent media coverage, while others are more obscure. Mākua Valley, an area of seventy-two hundred acres on the Wai'anae coast of O'ahu, was seized in 1941 under martial law and has since been a site of struggle between local residents, state authorities, and the U.S. military (Kelly and Aleck 1997). In 1983 the military, referring to the Native Hawaiians living on the beach as "squatters," persuaded then-Governor George Ariyoshi to evict them (Rohrer 1995). The common leeward coast practice of living on the beach in the summer (DeCambra 1997) was redefined by the military, developers, and the state as illegal. This highly publicized incident is but one eviction in an ongoing series of physical struggles over land. In the summer of 1997, the military catapulted Mākua into prominence again by proposing to land about seven hundred marines and their equipment at Mākua Beach on their way to exercises in the valley (Witty 1997: A-1, A-8). The announcement came just days after hundreds of people gathered at Mākua Beach to take part in scattering the ashes of Israel Kamakawiwoʻole, a cultural icon of Hawaiian music and an advocate of Hawaiian sovereignty. The military apparently failed to take a cue from the obvious significance of his death and burial at this place of Hawaiian spiritual meaning, even though he had lain in state at the Capitol where thousands lined up to bid him a tearful farewell preceding his televised funeral. The death of "Bruddah Iz" intensified Native Hawaiian convictions concerning the cultural importance of Mākua Beach. At a subsequent public meeting held by the Waiʻanae Neighborhood Board (which voted thirteen to zero against the proposed landing), Hawaiian activists promised unparalleled civil disobedience on the coast if the operation was carried out. While insisting on their possible future use of Mākua Beach, the military landed the marines elsewhere and ferried them into Mākua Valley for the exercises (Omandam 1997: A-1; Morse 1997: A-3).

Other conflicts in Mākua concern the "open detonation of surplus ordnance and open burn of toxic military and hospital wastes" (Rohrer 1995: 6) and the environmentalists' challenge to the military's destruction of the endangered Hawaiian tree snail's habitat (Rohrer 1995: 5). Army Secretary Togo West Jr. dismissed the protests by invoking the magical, self-justifying incantation, national security: the protests "must be weighed against the mission of the Army, which is to protect national security interests" (quoted in Rohrer 1995: 8). Yet it is unclear exactly what "weighing" is going on, when "national security" performs in the hegemonic discourse as a self-referential reassurance rather than as a contestable concept.

Lualualei, a large parcel in the Waiʻanae area, is another publicly contested space, with grassroots groups and, sometimes, state officials pressuring the federal government to return land and raising questions about the health risks of electromagnetic radiation from a nearby navy transmitter (Rohrer 1995: 13).[27] A somewhat less publicized land battle on the Windward coast of Oʻahu is the Kamaka family's court battle with the navy and marines to keep their 187-acre farm in the Waikāne Valley (Tummons 1992b). Sovereignty activists politicize the prevailing spatial arrangements by pointing out, for example, that the University of Hawaiʻi sits on ceded lands and that the popular Ala Moana Shopping Center occupies land from which Chinese, Japanese, and Hawaiian gardeners were dispossessed in the early 1920s.[28]

Environmental concerns arising from military land use are woven into sovereignty issues and are also sometimes voiced from more establishment-based sources. Former State Deputy Director for Environmental Health Bruce Anderson pointed to environmental dangers posed by military storage facilities in a 1993 front-page story in the *Honolulu Advertiser* (Yoshishige 1993). In 1994 the Environmental Protection Agency added Lualualei to the Federal Superfund National Priorities list, not because of the sixteen cases of childhood leukemia found among the population living near the controversial navy transmitter but because of the other eight hazardous waste sites found there (Waite 1990: 3).[29] Pearl Harbor also resides on the Superfund list of toxic waste sites, but the local press took pains to assure readers that a sizable oil spill there in 1996 posed no danger to the environment.[30] Like the periodic violent acts of military personnel toward their families or others, the military's violence to the land is generally treated in the mainstream media as unfortunate and separate incidents rather than as evidence of a damaging institutional presence.[31] It is in the Native Hawaiian sovereignty movement and some environmental organizations that the military's effects on local land and people are thematized as institutional practices rather than discrete events.

Perhaps the most well-known of the Native Hawaiian struggles with the U.S. military over land, and the most intimately connected with the growing sovereignty movement, concerns the island of Kahoʻolawe. Kahoʻolawe is the smallest of the eight main islands in the Hawaiian chain. Its twenty-eight thousand acres are home to a number of ancient archaeological sites as well as endangered plants and rare insects. Used for ranching prior to martial law, the island was taken over by the navy and used for bombing practice from 1940 to 1990. A grassroots organization called the Protect Kahoʻolawe ʻOhana began a vigorous campaign in 1975 to halt the bombing and return the island to the Hawaiian people. Although included in the National Register of Historical Places in 1981, Kahoʻolawe was bombed for another nine years after that year. In 1990 President Bush ordered a temporary halt to the bombing. Four years later Congress voted to stop the bombing permanently and to allocate funds for cleanup and restoration of the island (Rohrer 1995: 21). The recovery of Kahoʻolawe by the state has stimulated hopes for other successful struggles with the military for land: "As one activist reportedly said to another during the ceremony of return, 'One down, seven to go'" (Merrill 1994: 236).[32]

The singular successes of the Protect Kahoʻolawe 'Ohana and some other local land struggles take place against the dense background din of militarized discourses and the continuing immense presence of militarized institutions. There is no mention of Kahoʻolawe in the Fort DeRussy Army Museum. When Congressman Neil Abercrombie made modest suggestions about returning some of the military land at Bellows Air Field in Waimānalo to the state, he was subjected to an unabated flood of accusations about being antimilitary, even though he has consistently supported expansion of military housing and infrastructure in the state (Napier 1995: 114; interview with Abercrombie 1993). The disjuncture between continued public challenges to military land holdings from Native Hawaiian and environmental groups and the hysterical response to Congressman Abercrombie's mild criticisms of selected aspects of military land use suggests both that contestatory voices are audible and that the hegemonic views, as told in the Fort DeRussy narrative, are still firmly in control. The Native Hawaiian and, to a lesser extent, environmental groups have created vital public space for thorough critiques of the military's occupation of Hawaiʻi. In this sense the necessary space for facilitating the contending voices at the heart of a more democratic citizenship is being shaped. The continued democratic rewriting of Hawaiʻi's social spaces requires discourse that calls attention to the prevailing assumptions in order to make room for more critical perspectives. In Roxanne Doty's words, critical thinking must insist on calling attention to the performative in order to make the pedagogical able to be contested: "It is at this ambiguous margin that the double-writing of statecraft takes place and that the nation-states' identity, authority, and sovereignty are written and rewritten, but never finally fixed in a stable and unambiguous way" (Doty 1996a: 180).

Our aim in this chapter has been to contribute to a different peda-gogy, one that is able to call attention to the taken-for-granted *as* taken for granted, and to risk putting itself into question. We also wish to join with different kinds of performances, different relations between the grammar and rhetoric of political discourse, so that the production of meaning claims is more readily contestable, more open to difference. Challenges to the military's representations of Hawai'i are both vigorous and submerged, both persistent and mar-ginalized. Our goal is to demonstrate how, *in discourse,* this margin-alization is both accomplished and contested. By focusing attention on the inherent ambiguities in political narratives, and on the on-going projects of writing and rewriting order onto potentially resis-tant bodies, we hope to stimulate more heteroglossic practices of statecraft and more democratic notions of citizenship.

4

Remembering and Forgetting at Punchbowl National Cemetery

Locally known as Punchbowl Cemetery, the National Memorial Cemetery of the Pacific in Honolulu is the burial site for military personnel killed in the Pacific in World War II and the Korean and Vietnam conflicts, and for veterans of those wars. Visitors enter the cemetery, located in a hot, dry part of the city, by way of a winding road that climbs around the outside of the crater. As they ascend, the seaward side offers a handsome vista of the central part of the city. Night-blooming cereus and other succulents grow easily along the road, nearly obscuring the black lava and cinders of the crater. Passing through a wrought iron gate, the road opens onto the gently contoured terrain of the crater, which hosts an abrupt change into green vegetation. The expansive interior of the crater, more than one hundred acres of carefully mown grass, is landscaped with palms and monkey pod trees and many thousands of flat white rectangular headstones, traversed by gracefully curving drives (fig. 12). The city's dynamic profile of skyscrapers jostling one another for eminence in height is suddenly displaced by the appearance of this formal, quiet garden. Directly ahead of the gate at the back of the crater, several score of wide graceful white stone steps rise through the terraced Courts of the Missing to the crowning arc of the monument, a set of tiled maps depicting the Pacific War and concluding in a nondenominational chapel. Off to the far right sit the columbaria holding the ashes of other veterans.

Figure 12. Aerial view of Punchbowl National Cemetery. Courtesy Island Heritage Publishing.

While Punchbowl is internationally known as a site for the burial of soldiers (and sometimes for their next of kin as well), it is also a site for the production of the stories the state tells about why young men die in war. There are two distinct kinds of spaces produced at Punchbowl: the burial space, occupying most of the area with neat symmetrical rows of graves, and the space of the memorial, with its texts, maps, inscriptions, illustrations, instructions, and prayers.[1] At first glance, the second space appears to be full of words, while the first is starkly silent. But the orderly march of soldiers' graves inhabits a dense historical text, where "national narratives are written directly onto material soil" (Diller and Scofidio 1994: 28), while the signifying practices of the memorial are also riddled with silences. The "silent city" of the war dead offers a solemn surface where meaning appears as absolute (Robin 1995: 55). Yet death's presence is a potential reminder of other possible stories. Considerable interpretive work is necessary to produce and sustain the various possible memories of international carnage within the realm of the sacred. All roads in the cemetery lead to the memorial, where the official version of the wars is elaborated in art, architecture, and information. The space of the graves and the space of the memorial are mutually constitutive of one another, producing through their interactive imagery a set of stories that pacify death, sanitize war, and enable future wars to be thought.

Because sites for the commemoration of mass death have to struggle to represent their expected and endorsed image, familiar cultural figurations and reliable tropes are called into service to stabilize meaning. Representations of masculinity and femininity are useful recruits from Nietzsche's "mobile army of metaphors" (1956b: 180); they assist in bringing other possible remembrances to order around a single commanding memory, "an undifferentiated time of heroes, origins, and myth" (Nora 1989: 8). Images of masculinity and femininity do their familiar duty at Punchbowl, inviting unarticulated associations concerning appropriate order and necessary sacrifice. Both the memorial and the cemetery are tacitly gendered spaces, where inscriptions of masculinity and femininity are hidden in plain sight/site. Tracking "the often silent and hidden operations of gender" (Joan Scott quoted in Inglis 1989: 35) at Punchbowl leads into an implicitly masculine space, one that is planned, controlled, disciplined, orderly. The memorial tells a monoglossic, regulated story of external danger and national victory. (The Vietnam War is omitted from the story, although dead soldiers from Vietnam occupy many of the graves.) There is no disorder, no mystery; there are no unanswered questions, no loose ends. Feminine figures are incorporated into crucial supporting roles in the narrative of manly conquest, but there are no autonomous voices or spaces that could be coded female.

Alongside the pomp of militarized memories in the memorial is the circumstance of undifferentiated, singular representations of dead men in the cemetery. Control abounds: no animals, no stray saplings, no unapproved flowers. There is a great deal of open, quiet space in the cemetery, but there is nothing peaceful about it. So-called civilian cemeteries in Honolulu and elsewhere often reveal a certain acceptance of jumble, of differences in the size, scale, inscription, and tilt of headstones, the placement of trees, the arrangement of flowers. Community cemeteries often present many invitations to enter; they allow memory to grow. Punchbowl manufactures a predigested set of memories; like the dead men beneath the ground, the visitors are all expected to march to the same drummer.

THE POLITICAL ANATOMY OF DETAIL

The name *Punchbowl* was bestowed in the nineteenth century by an English-speaking migrant to Hawai'i; the name is not a complex metaphor. Geologists call it a truncated volcanic hill and think it was formed in one gigantic explosion of "superheated steam-charged mud and rock-stuff" up through the relatively soft coral plain that had formed over the volcano (MacCaughey 1916: 609). The winds that day were not the normal fresh northeast trades that deposited ash and muck in the characteristic elongated pattern on the southeast sides of the vents. Rather, it was a day of the more rare Kona winds that piled the ash and debris on the northwest side. Punchbowl's highest point is five hundred feet above sea level and it looks down over the city of Honolulu, the harbor, and all of the ocean beyond. Its bowl-like interior slopes gently downward.

In Hawaiian mythology, *Pele,* a female deity, is credited with the creation of several kinds of landforms, including volcanoes. Searching for a peaceful existence, she roamed O'ahu probing the earth in places, causing craters, including Punchbowl, to appear. When she later departed for Maui, she extinguished the fire of all of the volcanoes on O'ahu.[2] After her came the *menehune,* an ancient race of people commanded to live in the same area. They were skilled at building with stone. A few fish ponds and walls still visible today are attributed to them (Nogelmeier 1985: 25). Hawaiian legends also retell the genealogies of many of the chiefs who later inhabited O'ahu and the area now known as Punchbowl, including the famous Huanuikalala'ila'i in the twelfth or thirteenth century. Known as a wise leader, he was more interested in farming than in war. Under his stewardship, the area became known for its agriculture, pursued inside the crater as well as on its slopes. A photograph taken around 1920 shows an aqueduct system and small farms, some of which perhaps date back to that early time (Nogelmeier 1985: 30, 31).

Punchbowl's Hawaiian name was probably *Pūowaina.* In the Hawaiian understanding, space was not a static thing but came into existence through breath, life, and speech. In English, *Punchbowl* suggests a fixed container capturing that which lies within, while the spatial representations in Hawaiian are more fluid and dynamic. Hawaiian overflows with meaning, its place names carrying a rich freight of meanings, stories, poetic speech, jokes, and proverbs

(Nogelmeier 1985: 7–13). Translation from Hawaiian to English is consequently not a straight route and usually entails a loss of meanings.[3] With that caveat, it is possible to say that the most likely meaning of *Pūowaina* is "a poetic contraction of Pu'u-o-waiho-'ana, hill of placing, connoting the places of sacrifices atop the hill on the historically recorded altar of human sacrifices there" (Nogelmeier 1985: 16). The spindle of Hawaiian culture was *kapu,* the religious practices that drew and held gods and the highest-ranking chiefs together. *Kapu* produced the nuanced, thoroughly gendered spatial practices that constituted and maintained social relations and the "space of kings" (Lefebvre 1994: 308). Sacred space was reserved for the *ali'i* (ranking chiefs), some of whom were so sacred—that is, so directly descended from gods—that they created their own space. Queen Keopuolani was an example of a highest status *ali'i* whose person and shadow were sacred. *Ali'i* of this degree of sacredness sometimes refrained from venturing out except at night in consideration of the lengths to which others had to go to avoid violating the *kapu.*[4] In this rolling spatial imaginary, places vibrate with the interactive energies that constitute them.

Pūowaina was part of a complex of *heiau* (temples) in the area, which probably owed their origins to the ritual burning there of the bodies of those who violated the sacred laws or space. *Kahuna* (male priests) first put the transgressing men and women to death by drowning them in one of the salt ponds seaward of the crater. Then their bodies were taken up to *Pūowaina* to be burned on a large sacrificial stone where the volcanic explosion had created a chimney-like crevice providing a strong draft.[5] The meaning of this kind of sacrifice differs from that professed at the National Memorial Cemetery. While both are practices for maintaining order through violence and both orders are hierarchical, the Hawaiians "did not overcode their violence with strategic rationales" (Michael J. Shapiro 1997: 36). *Kapu* bound chiefs and commoners to one another in relations of mutual obligation: the commoners to feed the chiefs, the chiefs to intervene properly with the gods. *Kapu* violations interfered with the social order and those relations to the gods; the most serious ones required the death of the violator, a violence that observed and restored the order. There was nothing anonymous or impersonal about it.

Historical reconstructions such as these ruffle the still and hallowed representation of today's Punchbowl, reminding us that there have been other ways to decode the space. To read it is to turn away from the idea of space as transparent or as an empty container devoid of social history and relations. Spatial "realism" leaves unseen and unchallenged the social and political forces, including neocapitalism and the modern state, that produce modern hegemonic understandings of space. Alternative spatial imaginaries, such as those of Native Hawaiians and many other indigenous cultures, enact dynamic and interactive views of space, reminding us that "physical space has 'no reality' without the energy that is deployed within it" (Lefevbre 1994: 13). Indigenous spatial arrangements frequently "embody a communal reason for existence" (Brotherston 1992: 83); they map not passive places but active spaces of life. "Native cartography," Gordon Brotherston observes in his study of Mesoamerican texts, appeals "to a deeper notion of geography that does not exclude the cosmic movements of the sky, history, or even therapy" (1992: 90). Lynn Wilson observes in her ethnographic conversations in Belau a similar dynamic relation between the land and the ways of living on it, an energetic manyness, "always-two, never-one" (Wilson 1995: 2). "Who a person is, who a family is—it's the land," remarks Cita Morei, a resident of an area called Olngebang in Koror, a city of Belau (quoted in Wilson 1995: 69). Everyday spatial practices link labor and identity to a mobile set of relations between people and land, people and ocean, people and reef: "Each piece of land mentioned [in the family stories shared with Wilson] has a name and a clan associated with it, so the stories represent not only the movement of relatives from one place to another but of their associations and relationships with the people who lived there" (Wilson 1995: 79).

Similarly, in Hawai'i places were named by their relations to the movements of natural phenomena, people, and gods. Rocks and sharks could be invested with profoundly spiritual meanings. *Kalo*, the staple food, was understood as sharing ancestry with humans.[6] Winds, waves, beaches, reefs, mountains, cinder cones, and springs had names. Communal access to land, extensive sharing of food, and complex interactive responsibilities among people, land, and ocean were/are entailed in the concept of *'āina*, usually translated as "land," literally "that from which one eats" (Kameʻeleihiwa 1992: 9). Basic terms of human kinship reflect names of plants, water, and land: "The basic Hawaiian family unit is called the *'ohana*, a term derived from *'ohā,* root offshoot of the parent taro plant. The terms for alternative generations, *kupuna,* grandparent (as well as ancestor), and *mo'opuna,* grandchild (as well as descendant), echo the term *puna,* spring of life-giving water, a term symbolizing the assurance of human continuity and life itself" (Meinecke 1984: 92).

A "very complex and dynamic system" of land tenure preceded the midnineteenth century *māhele,* the "sweeping transformation from the centuries-old Hawaiian traditions of royal land tenure to the western practice of private land ownership" (Moffat and Fitzpatrick 1995: 11). In the traditional system, districts were divided into many *ahupua'a:*

> The *ahupua'a* were usually wedge-shaped sections of land that followed natural geographical boundaries, such as ridge lines and rivers, and ran from mountain to sea. A valley bounded by ridges on two or three sides, and by the sea on the fourth, would be a natural *ahupua'a.* The word *ahupua'a* means "pig altar" and was named for the stone altars with pig head carvings that marked the boundaries of each *ahupua'a.* Ideally, an *ahupua'a* would include within its borders all the materials required for sustenance—timber, thatching, and rope from the mountains, various crops from the uplands, kalo from the lowlands, and fish from the sea. All members of the society shared access to these life-giving necessities. (Kame'eleihiwa 1992: 27)

Metaphors for *'āina* include both mother and child, both elder and younger sibling (Kame'eleihiwa 1992: 32). Caring for the *'āina* was caring for and serving one another. Central to relationships among *ali'i* and commoners, ancestors and gods, *'āina* was a dynamic space that "was given from one person to another, but was never bought or sold" (Kame'eleihiwa 1992: 51).

Celebrated by foreign economic interests as a triumph of civilization and by missionaries as the only way to render the Hawaiians "industrious, moral and happy," the *māhele* eroded the inspirited spatial practices of the Native Hawaiians (R. C. Wyllie quoted in Kame'eleihiwa 1992: 202). Many commoners protested the sale of land to foreigners and the dominance of foreigners in politics through a massive petition movement to their chiefs (Kame'eleihiwa 1992: 329–40; Kealoha-Scullion 1995; Hall 1985). The experiential and spiritual energies inhabiting Hawaiian spaces were/are assaulted by the imposition of the western spatial imaginary, the rationalized grip of the grid: "When the surveyors of the *māhele* took to the field . . . [they often] failed to appreciate the sophistication and complexity of the ways in which Hawaiians functioned within their environment" (Moffat and Fitzpatrick 1995: 17). The straight lines and precise angles of the first surveyors eroded the lived complexities of Hawaiian spaces and paved the way for the full-on linear rationalization of space, much as the legal division and sale of land enabled its rapid widespread commodification. In their study of the first western surveyors, Riley Moffat and Gary Fitzpatrick sum up the differences in spatial imaginaries:

> The differences between the Hawaiian and Western systems of land tenure can be traced to very basic religious beliefs. The argument has been made that in Hawai'i, land belonged to the gods, and that the *ali'i'ai moku* was merely the steward of the land. In previous literature, therefore, the descriptions of the Hawaiian land tenure system err in stating that land "belonged" to the king. The Hawaiian concept of stewardship is quite different from Western practices, where "property rights" make land a commodity that individuals own and can do with as they please. (Moffat and Fitzpatrick 1995: 17)

The surveyors of the *māhele* did not impose their cartographic imagination on a static traditional system but, rather, onto a living web of negotiated spatial practices already manifesting the consequences of dying knowledge, dying relationships, dying people:

> People were moving away from the land, they were dying faster than they were being replaced, knowledge about the original land tenure system was disappearing, the *ali'i*—the principal landlords—were viewing land in a different economic sense, and the relationships between those who controlled land and those who worked it were becoming less harmonious. (Moffat and Fitzpatrick 1995: 20)

Contemporary Native Hawaiian political, economic, and cultural movements often reemphasize the dynamic interactive connections among people and places, articulating an indigenous spatial politics on which claims to sovereignty can rest (Riveira 1990–91; Trask 1993; O'Donnell 1996a).

The silent city of Punchbowl manifests the strategic control of grid-like spatial practices. It is codified space, "tied to the relations of production and to the 'order' which those relations impose, and hence to knowledge, to signs, to codes, and to 'frontal' relations"

(Lefebvre 1994: 33; also n. 16). Certainly all spaces are inhabited by codes, but codified space, following Henri Lefebvre, is monoglossic, unidirectional, and generally self-referential; it is an order that can be written on the land without reference to any other logic, as did Pierre-Charles L'Enfant in the design of Washington, D.C. Codified space carries a generative role in the creation and reproduction of modern hegemonic social order.

The removal by force of Queen Lili'uokalani by American businessmen and marines in 1893, followed by the American annexation of Hawai'i, was the rupture that began the most recent redefinition of *Pūowaina* as codified space. The spatial centering of the United States and the consequent flattening of the Hawaiian sphere have been brilliantly illustrated by R. Douglas K. Herman (1996). He traces the movement of the word *kapu* from its sacred world to its use as the secular guard of private property and the American social order.[7] Signs lettered KAPU began to appear, proclaiming ownership and forbidding entry to newly defined public/private space. The condensed abstractions of straight lines and legal boundaries articulated the practices of standardization, accumulation, and control imposed onto the contours of the land.

The transformation of the unique premodern space of *Pūowaina* into codified space did not take long; it is a metaphor for the colonization of Hawai'i itself. In the nineteenth century, cannons were mounted on the fluted rims of the crater by the nascent Kingdom of Hawai'i (which became a constitutional monarchy in 1840). Aimed defensively at the harbor, the guns became primarily markers of sovereign power, fired principally on ceremonial occasions. King Kalākaua, members of his court, and others later planted trees on the rim; his soldiers brought water for the trees up the hill in casks. Townspeople walked and drove their carriages on the road winding up the crater in order to enjoy the quiet, their picnics, and the splendid prospect of the city, the harbor, and the rest of the island. During the continual improvement of access to Punchbowl, the sacrifice rock was itself sacrificed to the construction of a large observation area. The external slopes of Punchbowl were settled by Portuguese families who grew a variety of crops and built homes; until the early 1960s Punchbowl was the starting point for the Portuguese Holy Ghost festival procession (*Honolulu Advertiser* 1994a: G-12).

The December 1941 attack on Pearl Harbor, demonstrating

forcefully that Hawai'i had become American soil,[8] was the event that brought the cemetery to life, as it were. Ironically, in November 1941, Congress had appropriated $50,000 to construct a national cemetery in Hawai'i. However, since the territorial government did not rush to donate the needed land for it, the idea languished until 1948, when the decision was made to institute it.[9] Movement was swift thereafter; on January 9, 1949, the first of the mass burials of about thirteen thousand American war dead took place.[10] These burials were closed to the public. Single, public burials were begun about six months later. The first was that of war correspondent Ernie Pyle. The names of another eighteen thousand dead, whose bodies were never recovered, are inscribed in the marble Courts of the Missing (Carlson 1990). Violence is both hidden and inscribed in the abstract rationality of the cemetery's development and the arrangement of its graves. They are laid out in evenly spaced rows with flat headstones incised with the names of the dead men (when known), their home states, an alphanumeric-coded cemetery location, a religious symbol (choices limited to Christian, Jewish, and, after much prodding, Buddhist), their rank and branch of service, their war, and their birth and death dates. Discipline, defined by Foucault as "a political anatomy of detail" (Foucault 1979: 139), brings codified military order into the potential chaos of grievings, ruptures, and recoveries that might otherwise erupt in other "disorderly" spatial practices of death.

When space is taken as a transparent medium, cemeteries are places where the bones and other remains of the dead are invisible and underfoot, present but absent, while the personal feelings and decorum of visitors are visible and socially regulated. An understanding of space as an active producer of the social order excavates the dead and politicizes their absence; it raises the question of *how* they are bones. There is nothing left to chance at the cemetery, maintained as "hallowed grounds and national shrine." The cemetery has become codified space, the space of scientists, planners, and technocratic subdividers, where the possible variety of spatial imaginaries is called to attention by the state. Codified space, "in thrall to both knowledge and power," produces the authoritative instrumental practices that engender "the silence of the 'users' of this space" (Lefebvre 1994: 50, 51). The abstract and self-referential codes of such space are among the disciplinary state's techniques of power, utilized for the management of populations more than the governing of citizens. Populations constitute economic and political data, interchangeable and subject to manipulation; they are thus a legitimate "field of intervention, and . . . an object of governmental techniques" (Shapiro 1994: 484). Population's "peculiar variables," Foucault writes, are produced "at the point where the characteristic movement of life [and death] and the specific effects of institutions intersected" (Foucault 1978: 25). Combining Lefebvre's sensitivity to the way the state's "rationality" organizes society with Foucault's concept of population management, we see the practices by which a "hypertrophied analytic intellect" (Lefebvre 1994: 308) puts the state into the identity-formation business (Shapiro 1994: 485). At Punchbowl, the dead, as a static population, have become state data. Augustus Saint-Gaudens captured this when he said of Arlington National Cemetery, "Nothing could be more impressive than the rank after rank of white stones, inconspicuous in themselves, covering the gentle wooded slopes and producing the desired effect of a vast army in its last resting place" (quoted in Jackson and Vergara 1989: 26).

Cemeteries can manifest other sorts of spatial arrangements. They can be a symbolic space that is dialogical and heteroglossic, calling up a rich and diverse texture of memories and narratives. Until the end of the nineteenth century, the common Anglo-American practice was for the friends and relatives to lay out the body and then bear it to the grave. Burial was generally in the symbolic space of a churchyard (Mitford 1978: 17), which permitted a variety of relationships to life and death.[11] In Hawai'i, the Cemetery Research Project (Purnell 1987) located the sites and tombstone inscriptions (including names, if any) of the rapidly disappearing burial places of the past. Such places were almost entirely the burial sites of Native Hawaiians and early plantation workers and their families, reproducing the exclusionary pattern of a century of colonial dominance. The cemeteries were found in backyards, caves, shopping malls, the heart of the city, and at the edge of the ocean. Graves were marked by stones varying from rough-hewn lava rocks to water-shaped ones clearly selected for their beauty. Some ethnicities buried money with the dead; others affixed a photographic likeness of the person to the rock or headstone. Languages varied: Ilocano, Tagalog, Chinese, Japanese, Hawaiian. The inscriptions often included names and birth and death dates, as well as parents' names and place of birth; sometimes they included the circumstances of death (a tidal wave, for instance). This is not the masculine space of abstraction and hierarchic order, but the symbolic space of networks of living beings, tied to identities and concrete lived experiences where loss was tangible and open to a variety of interpretations. Besides recovering the idea of a heteroglossic cemetery, the project has enabled Native Hawaiians to regain important genealogical history and to confirm family ownership of cemetery sites. Without such knowledge, Native Hawaiians can lose these gravesites to the codifying practices of planning experts and developers.

The spatial practices of the community-based, kin-oriented, incrementally ordered local cemeteries could be coded feminine, in contrast to Punchbowl's vertical, masculine order. The voice of fathers speaks loudly at Punchbowl. There is nothing accidental or overlooked about the policing of the tasks peculiar to managing the dead as data, from their interment to the maintenance of the cemetery as hallowed grounds, national shrine, and tourist attraction.[12] The burying of the initial thirteen thousand war dead echoed the impersonal procedures of the assembly line.[13] Army photos taken at the time show long lines of what appear to be metal rectangular boxes, looking very much like square-cornered steamer trunks. They were stacked two high, ready to be lowered by crane into mechanically dug trenches. Hundreds were buried each day to meet the "requirements for speed and efficiency and for respect due to America's war dead" (Carlson 1990: 34). Construction of the rest of the cemetery continued during these burials. It was formally dedicated on September 2, 1949, the fourth anniversary of V-J Day.

The state exercises extensive surveillance over the bodies of the estimated five million visitors a year.[14] At the base of the massive statue of Columbia, which looks out over the cemetery, a sign lists (in English and Japanese) the following prohibited activities:

Any form of sports or recreation . . .
Boisterous actions and disrespectful conduct
Smoking
Consumption of food or beverages
Bringing pets into the cemetery
Personal maintenance of graves
Failure to respond to announcements of the cemetery's closing
Failure to observe floral regulations [15]
Failure to observe traffic control signs

Also posted is the egregiously overbearing information that transgressors are subject to prosecution under both federal and state law. As with other attempts at total order, this one does not fully achieve the seamlessness at which it aims. Local people are sometimes seen tending the graves, laying personalized markers representing nonregulation memories. A local student recalls going to Punchbowl with well-stocked coolers for a family picnic at the gravesites of family members. This young woman, an officer in the reserves, found herself both moved and embarrassed by familial liberties taken with Punchbowl's rule-governed territory.[16]

For Punchbowl to succeed, that is, to promote itself as codified space, its "'users' [must] spontaneously turn themselves . . . their lived experience and their bodies into abstractions, too" (Lefebvre 1994: 93). Punchbowl's success requires visitors to recognize themselves in what Lauren Berlant has called "the self abstraction of citizenship"; this abstraction teaches adults to "forget or render as impractical, naive, or childish their utopian political abstractions, in order to be politically happy and economically functional" (1993: 399). The state's story and logic of death maintain war as the dominant imaginary. A citizenship that insists on interrogating the practices of war, or questioning its necessity, is not welcome at Punchbowl. Yet the practices of critical citizenship can dislodge war's self-abstraction and invite other imaginings of death and the state.

HEGEL COMES TO HONOLULU

Hegelian tropes of history, nature, death, and the state figure prominently in the narrative practices of the memorial.[17] Perhaps it is indicative of G. W. F. Hegel's continuing influence on national narratives of violence and progress that this memorial, commissioned by the American Battle Monuments Commission and built by a coalition of architectural firms from San Francisco and Honolulu, so neatly reflects Hegel's story. Or perhaps it is only the perverse reading practices of political theorists that would make it so. At any rate, it seemed to us that we invited Hegel to the memorial, only to find that he had been there before us, providing conceptual tools for successfully shaping memory and grief and reverence into contours acceptable to the state. Hegel, the Battle Monuments Commission, and the modern nation-state all speak the same language, a language that projects a grand national destiny onto the future and then reads that destiny back into the past to provide a narrative of legitimacy and reassurance.

Entering an ongoing set of debates about revolution, the state, and social change, Hegel articulated a pivotal analysis of the relation between history and nature (or, in a more Hegelian tone of solemnity and grandeur, History and Nature). Hegel's take on the agonistic dynamic of human history as the ever enlarging arena of liberty has become an anchor for many of the stories that nation-states tell themselves about who they are, where they are going, and where they have been. For Hegel, the world of Nature is organized and expressed in the lives of women, children, and families—a world with its own ethical practices, which Hegel calls natural, but a world outside the grand dialectic of historical evolution and emergent reason, which has its home in the nation-state. Similarly, indigenous peoples from outside Europe and Asia also fall into the realm of Nature, lacking both the written language and the hierarchical institutions of state that provide the setting for the emergence of reason, self-reflection, and freedom. In contrast to the relatively static world of Nature (although for Hegel motion and change are everywhere) is the telos of History, a temporal stage across which march men, armies, and governments in a necessarily painful but ultimately successful move toward freedom and fulfillment (1975: 124–51).[18]

War, for Hegel, is a necessary part of this march. Hegel's version of war talk is not the garden-variety glorification of one nation at the expense of others, but a more pervasive ontological need for nations to affirm themselves through struggle with other nations. Like individuals, nations must periodically engage in the "negation" of hostile others in order to achieve self-recognition. The nation's identity is dependent on violence to produce its own coherence. One must have an antagonistic other to oppose in order to be an autonomous and free entity—at both the level of the individual (man) and of the state. To Hegel, the courage and sacrifice that states celebrate in war memorials do not mark a moment of individual bravery but an opportunity for the individual to be in solidarity with the nation: "The true courage of civilized nations is readiness to sacrifice in the service of the state, so that the individual counts as only one amongst many. The important thing here is not personal mettle but aligning oneself with the universal" (1967: 296). Hegel justifies war at an ontological level, as a violent necessity for both the citizen and the state "to reproduce or maintain the coherence of the body politic as a whole" (Michael J. Shapiro 1997: 45).

Intertwined with Hegel's categories of history and geography is a ruthlessly gendered imagery invoking martial desire, maternal love, and filial sacrifice. The object of violence in war, the enemy, is also an object of desire, bringing the self/state back to itself by way of confrontation. The labor of women in birthing families is necessary to produce the soldiers, and the labor of soldiers in dying is necessary to produce the state. Hegel's codes of history and nature intermingle with spartan stories of love and loss to produce a war-friendly monologue. Hegel contributes to the cultural repertoire of available narratives that enable the state to tell its story to the millions of visitors to Punchbowl, to replace possible images of torn bodies or frightened young men (or, for that matter, dead civilians or war-profiteering companies or devastated indigenous cultures) by making contending narratives unthinkable.

The bulk of the monument is taken up with a series of large mosaic panels illustrating selected events from World War II and the Korean War. Several paragraphs of written text accompany each panel, recounting the salient history of heroic battles, critical troop maneuvers, encounters with The Enemy, necessary sacrifices, temporary setbacks, and ultimate victory. These texts in Hegelian terms mark History: the domain of reason, men, and the state. Told from the implicit point of view of impending Allied victory, the telos of the story marshals all messiness, contingency, and uncertainty into a reassuring narrative of necessity. In this telling, the war, like the death it indiscriminately imposed and selectively recounts, was rational and orderly; it went largely according to plan. "Even before the amphibious assaults upon the Gilbert Islands were launched," one panel intones,

> the next stop in the Central Pacific advance, the air attack upon the Marshall Islands was initiated. As early as November 1943 aircraft of the Seventh Airforce and carrier-based airplanes of the Fifth Fleet started the preliminary bombardment of the Marshalls; in December and January the attacks were progressively intensified.

The intense detail and precision of the maps help to codify and regularize the war, to make it rational and strategic. There are arrows and markers depicting the planned movement of troops and equipment, the unfolding of strategies and tactics, the ultimate accomplishments (ours) and failures (theirs). The panels call on what

Lefebvre calls "the illusion of transparency . . . a view of space as innocent, as free of traps or secret places" (1994: 27–28). The maps and inscriptions simply Tell It Like It Was, eliciting excited recognition from at least some of the veterans touring the monument: "That's a very good map, it gives a good picture. . . . I've never seen so much detail in one map."[19] The crispness of the military dating system supplements the tone of authority and calculation:

> Exploiting their successful attack upon Pearl Harbor on 7 December 1941, the Japanese struck at American, British, Chinese and Dutch territories. The US forces, instantly upon the defensive, nevertheless determined to hold open the line of communication to Australia, to aid in its defense, and to regain her status in the Philippines.

The story starts here: no linkages to previous events or practices are made, no complex beginnings are recognized in global political economies. In the beginning was The Attack. All else follows, with offensive vices (theirs) and defensive virtues (ours) firmly fixed in place. War is depicted as a necessary reaction to a sudden external crisis—not, as Hobbes would have it, a pervasive inclination.[20]

Even when the plans seem to fail, the bad news is made into good news. The retreat of U.S. troops from Korea in November 1950 is rewritten as a technical victory:

> Surmounting heavy odds, bitterly cold weather, and rugged terrain, the marines and soldiers fought their way to Hungnam where together with other troops in NE Korea they were evacuated by sea and air to South Korea. Land and carrier-based aircraft and supporting naval groups proved invaluable in the redeployment.

And it all turned out right in the end: the armistice ending the Korean War is described as "bringing to a successful conclusion the United States defense of the Republic of Korea against the communist invaders."

The memorial tells a pious narrative (Shapiro 1988: 55) in which national victory is deserved and national purity is preserved. Their killing is irrational, unexplainable, misguided: for example, "the enemy reacted violently." Our killing is laudable and necessary: for example, "[the United States made] persistent and heroic efforts." Generous use of the first-person plural to describe U.S. troops conscripts visitors who are from the United States into the story: "Our ground troops" performed heroically; the American reader is part of the "us," which is America; therefore, the reader can make a semiotic slide into the ranks of noble soldiers and necessary battles.[21] Occasionally, the visiting reader is pressed into more active agreement with the worthy goals, necessary sacrifices, and efficacious planning of the armed forces. The Battle of Iwo Jima is described as one in which "we" met "a determined resistance conducted by a well-trained, well-equipped enemy." But it was worth it: "Undoubtedly, the numbers of Americans whose lives were saved by the operation of this air base exceeded the number lost in its capture." In the memory of the memorial, the war ends at Iwo Jima; there is no representation of Hiroshima or Nagaski. The memorial orchestrates a deafening silence about the atomic destruction of these cities; no account of massive civilian destruction interrupts the saga of loyal "brothers-in-arms" confronting an armed professional enemy. In the cunning of national reason, things work out right in the end.

Hegel's hand is visible not only in the dominant narrative of states and armies sweeping across the stage of History, but also in the supporting subtext of the memorial. Here, History is supported and confirmed by Nature, the static and wordless entities that exist outside History, in the atemporal stasis that is both necessary to and disregarded by those who make History, as well as those who tell it. Beneath the main mosaic panels runs a series of depictions of plants, animals, Pacific Islanders ("natives" to Hegel), and women. They are unaccompanied by written text. These are the elements that Hegel located outside of History because they have not yet manifested the potentiality of reason; they do not require investigation (Hegel 1975: 134). The smaller panels offer a kind of pictoral *National Geographic* to the dominant narrative's *Time-Life* documentary sagas. The mosaics feature tropical fish and sea horses; tropical birds and alligators amid flowers; a white arctic fox; a tangle of colorful tropical flowers; three albatross, one sitting on a nest, in front of a small body of water and a distant mountain; three dolphins leaping and diving in the water; dark-skinned islanders fishing and weaving in a peaceful lagoon; a Japanese garden, featuring two red-roofed pagodas, a graceful bridge across a small lagoon, and a woman in a kimono. The "natives" and the woman are as wordless and timeless as the flora and fauna, equally decorative, offering raw material to be worked on by history, but they themselves are not making that history. The natives, it seems, are happy, and the woman docile. We know the natives are happy because they are brown, they are wearing brief red garments (a mosaic version of a *malo* or loincloth) and flowers in their hair, and they are busy doing the simple, wholesome tasks they do when there is no war. We know the woman is docile because she is sitting quietly in a manicured garden, clad in traditional kimono, obi, and head ornament, her hands folded in her lap, kneeling, her head bowed.

Hegel's story, like the memorial's speeches and silences, put women and "natives" outside History, outside the dialectic of struggle through which fully human consciousness emerges. Human consciousness and human history, domains reserved for western men, are represented in terms of an idealized military encounter in which domination, violent struggle for recognition, and the willingness to die rather than submit are the hallmarks of emergent human, implicitly masculine, identity. The manly struggle for identity between individuals and nations legitimates and naturalizes war, conducted within what Hélène Cixous calls Hegel's "Empire of the Selfsame" (Cixous and Clément 1986: 79). The Other to this Same is, in part, Nature, represented within civilization by blood ties, reproduction, and death; women and families are stuck in nature, at best preserving "divine laws" of home and hearth, but never becoming bearers of reason or actors on the world stage (Hegel 1967a: 478). The imagery of nature in the silent subtext of the memorial echoes Hegel's representations of women, kinship, animals, and "primitives" as lesser, but still necessary, counterpoints to western masculine reason and desire. Happy natives and docile women are useful to the Hegelian narrative of the memorial in that they anchor these nonwestern and nonmasculine lives in the passive arena of the acted-on, while conferring both legitimacy and virtue onto the bold and necessary actions of the white men's armies on the world stage. Woman/native/plant/animal—the silent other that supports History without intruding on it, the blank page on which the western male pen is writing.

It takes some digging to excavate any other set of memories about the war in the Pacific. Interviews with Pacific Islanders tell a rather different story from that found in the mosaic image of peaceful, exotic natives in harmony with nature (White and Lindstrom 1989). Similarly, recollections of Japanese women, and work by feminist historians on the domestic impact of the war in Japan, also offer memories that erode the peaceful, passive image of the Japanese woman in the garden (Miyake 1991; Kanda 1989; Mackie 1988; Chuzo 1984). The impact of the war on the people of the Pacific varied widely. The fortunes of the Islanders were often tied closely to those powers that had colonized them before the war, usually Japan or European countries. In the areas in Melanesia colonized by Britain, male Islanders were often already organized militarily, and they took part in the war on the side of the Allies, fighting or coast watching. Many more joined (or were conscripted) into wartime labor corps for Allied military units. Remarkably few Melanesians were killed during the Pacific campaign, although village populations were often forced to move to other parts of their islands or, more rarely and with more serious consequences, to other islands. All this movement had a serious impact on social structures and material life. Able-bodied men were gone for long periods of time. Gardens were abandoned when villagers had to move; cultivating and living in new areas, under difficult conditions, became the burden of women, children, and older and infirm villagers (White and Lindstrom 1989: 3–40).

Significantly, the presence of American soldiers in large numbers gave Islanders a quick, powerful, and dramatic comparison to their colonial bosses for whom the color bar was a way of life. In a culture where status was marked by the ability to distribute goods, the willingness of American soldiers to share both their food and other goods with the male workforce held great symbolic importance. (In many cases, after the war's end colonial administrators took away from the Islanders what they had been given by the Americans.) Similarly, American soldiers repeatedly "violated the symbolic boundaries of [the colonial] inequality" by expressing their view that the Islanders ought not be called "boys" (White and Lindstrom 1989: 18). Such egalitarian gestures by the troops also violated general orders issued by their own commanders at a time when the American military was still strongly racist in its practices.

Life in several Micronesian islands was much more difficult and lethal. Men were conscripted for heavy labor, building Japanese fortifications, airfields, and depots. Once the supply lines to Pohnpei (Ponape) and Belau (Palau) had been cut, the Japanese garrisons on those islands were forced to live off the land, which meant they took food and domestic animals from villagers. Troops and villagers alike succumbed to starvation. Besides the threat of famine, Micronesians faced the dangers of severe American bombing raids on Japanese military targets (Nero 1989: 117–47; Ishtar 1994). The happy fishermen represented in the memorial's nature preserve might have had any number of different encounters with the personnel and practices of the occupying armies and would probably have many different stories to tell, but they were very unlikely to remain untouched or inactive.

In equally complicated ways, the war changed Japanese women's lives.[22] Enduring the enormous hardships of the war, they negotiated themselves and their families through dire food shortages, inadequate rationing systems, and an unreliable black market. Along with the elderly and the very young, women became the major providers of rural labor. They also entered the paid labor force in unprecedented numbers and in many cases challenged the traditional power relations within their patriarchal families (Miyake 1991: 293–95). The postwar legal reforms granting equal rights to women, usually seen as a gift from the enlightened Allied Occupation, were the fulfillment of the agenda of a vigorous prewar suffrage movement (Mackie 1988: 59). The active lives of women and "natives," their everyday social patterns in their particular places and times, are undetectable in the memorial's codified space. Punchbowl's appropriation of the symbolic spaces salient for Japanese women and Pacific Island peoples—the aesthetics of Japanese gardens, the spirituality of islands and oceans—turns these deeply symbolic, culturally embedded loci of meaning into snapshots of the exotic/erotic Other. Codified spaces, converted into the mosaic version of a tourist postcard, recruit symbolic spaces (Liggett 1994: 16). Completely absent from the memorial are the Japanese women, children, and men killed by the incendiary bombings of Japanese cities and the final atomic attacks. Tokyo, Nagoya, Osaka, Kobe, and dozens of other major cities in Japan were burned, in addition to the better known civilian slaughter at Hiroshima and Nagasaki (*Japan at War* 1980). While Japanese women might well have many different stories to tell about World War II, it is unlikely that many of the women would recognize themselves in the quiet, undisturbed figure kneeling in the garden. Once intersected by the voices of Pacific Islanders and Japanese women, the memorial's implicit claims to universality begin to look instead like very selective and self-interested accounts. "Nature" turns out to have its own ambiguous and constitutive history, if only one knows where to look.

Gendered codifications of national identity make a somewhat different sort of appearance in the inscriptions on the memorial referring directly to the dead and rotting bodies in the ground. Female bodies and feminine codes predominantly mark the death of men. At the top of the seventy massive steps leading up to the memorial, a gigantic thirty-feet-tall figure of Columbia, holding a laurel branch, looms in the visitor's gaze (fig. 13). Below this icon of a navy carrier's prow is engraved the spartan platitude said to have been offered by President Abraham Lincoln to a woman whose five sons were killed in the Civil War: "The solemn pride that must be yours to have laid so costly a sacrifice upon the altar of freedom." The United States is nearly always referred to as "she"; the dead soldiers are frequently referred to as "her sons." The memorial offers a tribute from a proud and loving mother to her worthy sons. Riding on Lincoln's rhetorical gesture, Mrs. Bixby, the bereaved mother of the five murdered boys, becomes The Mother/The Nation. Her sons become America's sons. She and, through her, the rest of us who loved or may in the future love the men who die in wars, are recruited into the state's instrumental relation to its citizens. Questions that grieving families might raise about the conduct of military operations, the wisdom of martial decisions, or even the legitimacy of the state and its war-making powers—these are all silenced by the reassuring image of the loving mother and her virtuous sons.

While the United States is nearly always memorialized in the feminine, The Enemy, in contrast, is always "he," always the masculine third-person singular. This linguistic marker locates The Enemy as a single, unified, ominous force; there are no pictures suggesting that The Enemy might be people. Only bunkers, ships, bombs, and abstract troop movements are portrayed. Individual U.S. soldiers are shown slumped in exhaustion, acting bravely in battle, and being kind to children, while The Enemy remains faceless and disembodied. The kneeling woman in the garden, while she could be interpreted in many different ways, does not disrupt the masculine unity of the Japanese soldier/Other; the Enemy, while outside the domain of martial virtue (being, after all, the bad guys) is still inside the larger domain of History (being, after all, still soldiers, still formidable, still men). As we read it, the pacific garden scene marks not the face of the enemy but the meek feminine outside, against which the bold masculine inside can be measured.

The stairs rise between two sets of courts containing giant marble

Figure 13. Steps rising up through the Courts of the Missing to the statue of Columbia at Punchbowl, with the rim of the crater visible in the background. Photo by Kathy E. Ferguson.

rectangles bearing the names of dead service personnel whose bodies were not recovered. At the bottom of these Courts of the Missing, the first wall sports two female angels, with generous, pointed breasts, holding a depiction of the obverse side of the great seal of the United States, which reads: "This memorial has been erected by the United States of America in proud and grateful memory of her soldiers, sailors, marines and airmen who laid down their lives in all quarters of the earth that other peoples might be freed from oppression." Another wall offers the following sexual speculations:

> In World War II 360,845 Americans gave their lives in the service of their country. They faced the foe as they drew near him in the strength of their manhood and when the shock of battle came they in a moment of time at the climax of their lives were swept away from a world filled for their dying eyes not with terror but with glory.

War is understood as a shocking moment, not a sustained disposition. Erotic fulfillment and national glory mix nicely; they were clearly having the (last) time of their lives.

While Hegel's narrative of History and Nature resonates loudly in the stories that Punchbowl tells, the memorial betrays Hegel in the end. For Hegel nothing is static; history always encounters nature again. The panels at the memorial give no quarter to the dialectic. There is no motion among their categories, none of the turbulent, organic struggle and reconciliation that accompanies the emergence of Hegel's *Geist* in history. Hegel, for all his categorizing, lets things move; within the telos of ultimate reconciliation, things are always mutating into their opposites. Parting company with Hegel, Punchbowl pins things down once and for all in the service of state-sponsored memories and "the sterile space of men" (Lefebvre 1994: 380).

Also, Hegel blurts out some of the colonial pretensions that the memorial politely evades. The memorial's representation of indigenous society is discrete and prim, a single panel of happy natives involved in timeless practices, a small marker of the feminized Primitive to contrast to the world-historical sweep of the Modern. Hegel, in contrast, is virtually pornographic in his eager gaze on his Primitive Other, which in Hegel's case is usually Africa or America rather than the Pacific Islands. Africa, says Hegel, is outside History because it has no written language (oral narratives do not count) (1975: 136), no political constitution, and no foreign relations with other states (i.e., with Europe) (1975: 138). There is no point in learning about the "savages" because, like families, they are outside of the fully ethical self-awareness through which the story of the emergence of freedom is told:

> The only appropriate and worthy method of philosophical investigation is to take up history at that point where rationality begins to manifest itself in worldly existence—i.e., not where it is still a mere potentiality in itself but where it is in a position to express itself in consciousness, volition, and action. The inorganic existence of the spirit or of freedom—i.e., unconscious indifference (whether savage or mild in temper) towards laws in general, or, if we prefer to call it so, the perfection of innocence—is not itself an object of history. (Hegel 1975: 134)

Having declared Africa outside the field of worthy investigation, his animated prose in *The Philosophy of World History* reports every strange account of native behavior with unabashed glee. Paralleling the missionaries' condemning gaze on Hawai'i, Hegel judges colonized cultures as "culturally inferior" because they were destroyed

(1975: 163): "They allow themselves to be shot down in thousands in their wars with the Europeans," proving that they do not value life (1975: 185). Lacking true culture, "wrapped in the dark mantle of night" (1975: 174), they routinely "eat human flesh," are "prone to fanaticism" (1975: 188) (unlike Europe, of course), are ruled by vile and lascivious leaders (1975: 189), are indifferent toward their children (1975: 185–86), and generally live lives of "sensuous arbitrariness" that are "incapable of any development or culture" (1975: 190). More remarkable than the contents of Hegel's speculations on his Primitive Other is his tone: his prose is excited, agitated, self-important, voyeuristic. Africa is a feminized spectacle, and watching is titillating; Hegel is horrified but cannot stop looking. He gobbles up every questionable scrap of information that comes his way.[23] He returns to the subject in the 1826–27 *Additions* to the text, noting that Africans slaughter each other in battle (1975: 219) (unlike Europeans, of course), routinely eat each other (1975: 220), and are quite contented as slaves (1975: 219). "But enough of this primitive and savage condition of man," exclaims Hegel in the last sentence of the *Additions* (1975: 220).

But Hegel cannot get enough of the barbaric, feminized Other who confirms his civilized, masculine self. Perhaps his pornographic gaze on the Primitive Other opens up one of the silences in the Punchbowl Memorial, where the texts can work up some erotic fascination with the glorious, mortal embrace of The Enemy but are quite demure about the sexual politics of encounters with indigenous cultures. Hegel's lascivious enjoyments are a bit more honest, perhaps, than the memorial's sedate colonialisms. Hegel and Punchbowl share the common western encodement of the dark-skinned Other as feminized, lesser—the static, natural outside against which the dynamic, civilized, masculine inside can make its appearance and do its work. The state can tell its stories about necessary and successful wars more readily in a discursive space that puts potentially critical voices outside the arena of reason, making them disloyal to civilization. While the Punchbowl Memorial declines to follow Hegel in his active sexualizing of indigenous peoples, the narratives at Fort DeRussy and other sites in Hawai'i are less reluctant to sexualize the Pacific Island other, more eager to impose the military's masculine self-understanding on the feminine bodies of Island women and Island land.[24] Concerning the gendered practices of colonialism, Hegel hides less, while the memorial rests on an orchestration of contested silences, enabled by the ruthless codification of space, about what was there before and how it came to be lost.

In the small chapel at the Punchbowl Memorial an inscription reads, "In proud remembrance of the achievements of her sons and in humble tribute to their sacrifices this memorial has been erected by the United States of America." This narrative of pride, love, and sacrifice gives comfort to actual mothers, fathers, wives, children, and friends whose beloved have died or will die in war. It also gives a great deal of comfort to the state, whose hand in causing war and sending men to fight it is thoroughly mystified.

The memory practices manifested at Punchbowl are anything but humble. The hubris of state power is hidden in plain sight: the dead soldiers "gave their lives" (no one took them), and now they "sleep side by side" (a sort of long, fraternal nap). Punchbowl shows the state at work producing the comforting fictions of necessity and freedom: the men freely chose to do what had to be done (rather than, say, being forced or constrained to do what might be questionable or unnecessary).

An information sign near the towering statue of Columbia explains the mission given to the creators of the memorial:

> In formulating the design of the cemetery memorials, the architect was limited only by budget constraints and the requirement that each memorial embody three basic features: a small devotional chapel, provision for the inscription of the names and particulars of the Missing in Action of the region, and a graphic record in permanent form of the achievements and sacrifices of the United States Armed Forces.

The designers of the memorial were under orders to make war into an achievement and death into a freely chosen personal sacrifice for the greater good. It takes a powerful set of narrative strategies to pull this off. The memorial's architects rose to the challenge by mobilizing a readily available set of narrative strategies about masculine reason, dark feminine otherness, filial loyalty, and maternal sacrifice—a rich, potent, hegemonic, but still contestable set of cultural insistences. Even the most seamless productions of reassurance and renewal are vulnerable to erosion from intrusive waters and persevering unexpected winds.

5

Seeing as Believing at the Arizona *Memorial*

Hawaiian Sam Ka'ai's spatial imaginary of the continental United States is as an *ocean* of land containing "strange islands" such as Walla Walla, Chattanooga, Houston, and Detroit. He contrasts these with his "cities of the Pacific": Rapa Nui, Vavau, Moloka'i, Tongatapu, O'ahu, Aotearoa. Where mainland dwellers solve problems with "longer roads and percentages," island dwellers walk on the water, "part of a voyaging group who . . . knitted themselves into the elements of their environment" (Ka'ai 1987). An island, he says, is a precious, small thing that cannot be totally used. A postage stamp where dwelling can only occur on the white perforated edges. Where rainwater trickles down on the colored part. Where, with the longest river fifteen miles in length, to think of water as a resource is "laughable."

He then asks us to imagine many visitors to this tiny, vulnerable place. Not your usual visitors, but a million angels who "along with all the high qualities, positiveness and smiles they would bring—the flapping of two million wings would spread a lot of feathers around. And you have to ask yourself if you like feathers mixed with the leaves and other things. Even the feathers of angels can become a foreign body introduced into your area" (Ka'ai 1987).

In this gentlest of metaphors Ka'ai has suggested some spatial presumptions of the practices of tourism. In his soft way, he wonders at the angels' bold sense of entitlement that legitimates for them the noise and cyclonic winds produced by their flapping wings and their imperviousness to the costs of the perpetual shower of moulted, ragged, sometimes broken, perhaps muddy feathers raining down. Although he does not offer us metaphors of air traffic control, surely this is necessary with two million moving wings; they must be directed from vector to vector, spatially and temporally dispersed, for not all the angels can dance on the head of a pin at one time.

It is not difficult to imagine that Ka'ai's angels are tourists or that a million of them refers to mass tourism, a "phenomenon of the twentieth century" (Diller and Scofidio 1994: 10). According to one estimate, 429 million persons vacationed in another country in 1990, spending an estimated $249 billion (Diller and Scofidio 1994: 10). As a set of social practices, mass tourism represents significant economic and political stakes for many places. Geographic regions, political units and promoters, and the tourist industry all invest a great deal in developing facilities and strategies for attracting and increasing the number of tourists to their areas. In the language of tourism, "visitor destinations" are excavated, restored, built, covered over, and/or landscaped in order to meet the expectations of the greatest number of tourists (Diller and Scofidio 1994: 10–11). Similarly, the expectations of tourists are developed through the cultural equivalents of bulldozer and crane operators, landscapers, architects, construction workers, land-use planners, banks, and realtors. These equivalents are the cultural forces and practices that form our identities and construct our mythologies (Diller and Scofidio 1994: 11).

It is a given that tourists go places to see things, for, as we say, seeing is believing. But *how* they see, as many have proposed, is a cultural production, not an act of individual raw vision.[1] No look is innocent; it is tied to expectations. This means that air traffic control for the six million angels[2] who annually visit the narrow edges of Hawai'i operates through a widely shared perspective combining the pleasure of looking with the assumption that the world is laid out before the eye/I as a picture.[3] The world as exhibition corresponds with the modern package marketed by tourist agencies. Kenneth Little emphasizes the erotic pleasure of the tourist gaze (Little 1991: 154–55), as well as the sense of control over the viewed objects produced by the "scopic regime" (Jay 1988). These entitlements explain much about the feathers and vortices; they are not in the frames because the culturally shaped tourist perspective is not reflexive; the visual practice of these modern hunters and gatherers is capture and control.

Tourists arriving in Honolulu after the five-hour flight from the West Coast are understandably ready to see almost anything. Depending on their aircraft's approach, they might see several other islands of the archipelago, or the white beaches and blue ocean of Waikīkī, but there is a special eagerness to glimpse Pearl Harbor as the plane comes in to land. The world as picture is a sight and relationship for which they are ready. But Hawai'i is a paradoxical world encompassing simultaneously the exotic register of desire—the carnival pleasures of beaches, tanned bodies, mai tais, palm trees, blue skies, and verdant lushness—and the register that annually calls a million and a half of them to the site of a major military disaster of World War II. Both landscapes of desire have been reinvented and managed, not duplicitously but as consequences of their "(re)creation as places of memory" (Diller and Scofidio 1994: 12). Such sites must be able to accommodate those who would visit. So what was once the original site is marked as "the real" or "the site," while the context—its spatial surround—is quite changed. A site must be marked in order to be apprehended as a site. Speaking of tourists as the "unsung armies of semiotics," Jonathan Culler explains their semiotic work as the exploration of "the relation of a sign to its markers" (1988: 165). Semiotically, the troops' efforts to see "the real" are always doomed to defeat by the iron law of the floating signifier: "the authentic" must be marked to be properly read.

American visitors may, in a sense that is important to them, "see" Pearl Harbor for the first time, but their reading of it was shaped earlier in their lives by the boundary-maintaining, identity-forming work of our national and economic narratives. Like tourists arriving in Kenya for a safari, they arrive in Hawai'i "with a perspective and a story in mind and try to find scenes that resemble these prior images that evoke recognition and an easy sense of familiarity" (Little 1991: 156). Pearl Harbor and the *Arizona* Memorial are easy pickings because they release narratives in which American self-recognitions are embedded. As Americans, tourists have absorbed the cultural and national mythologies surrounding the Pearl Harbor "sneak attack" on a sleepy Sunday morning on December 7, 1941. The details of the rest of the attack may be vaguer: that all eight battleships of the Pacific Fleet were hit, four sank, and the others suffered serious damage; that the USS *Arizona* sank at its moorings ten minutes after it took a bomb hit on an ammunition magazine and carried with it 1,177 crewmen, most of whom were trapped below decks; that human casualties of the raid ran to more than three thousand (Iles 1978: 11); that the fleet might have faced a total disaster had two aircraft carriers en route not been delayed by high seas. (Six months later, these carriers were instrumental in defeating the Japanese fleet at the Battle of Midway.) However shaky their grasp of details, most visitors are familiar with the narrative's end—the unconditional surrender by Japan in 1945 on the deck of the USS *Missouri*—even though an increasing number of them were born well after that event.[4]

This is not the only narrative possible or available, and the relationship between Pearl Harbor and the *Arizona* is more complex than the chronology of attack, victory, and dedication suggests. In "Discourse in the Novel," Mikhail Bakhtin fluffs up the pillow of language, conventionally flattened by the view of it as a "system of abstract grammatical categories" (1981: 271), into a plump cushion of views. For him, language hosts an ongoing contest between centripetal forces aiming at a centralized language and centrifugal forces that foster a heteroglossia. The familiar story of the attack is, in Bakhtin's terms, a "unitary language" (1981: 270), a unified and centralized verbal and ideological world. Yet the victory of centralizing forces over the contending voices is not the same thing as an unconditional surrender:

> At any given moment of its evolution, language is stratified not only into linguistic dialects in the strict sense of the word . . . but also—and for us this is the essential point—into languages that are socio-ideological: languages of social groups, "professional and generic" languages, languages of generations, and so forth. (1981: 271–72)

Virtually every move that tourists make in order to "see" Pearl Harbor and the *Arizona* Memorial interpellates them as consumers/citizens. They arrive in Hawai'i as part of a "voyaging group . . . knitted . . . into the elements of [an] environment" (Ka'ai 1987) of global capitalism, international politics, and the modern technology of jets. As air cargo, they are administered through management techniques (assigned seats in densely packed cabins) and pacification strategies (movies and drinks) that address them as consumers. Many are greeted at the airport by (to the tourists) "exotic-looking" young men and women wearing their workaday uniforms of "native dress," and given flower leis whose "tropical" colors might have been enhanced by dyes or spray paint. When tourists leave the baggage carousels as Arrivals, they are prepared to see it all and to photograph most of it. Monarchs of all they survey,[5] their realm is one that Culler (1988: 153) notes is "more imperiously an array of places one might visit than it is a configuration of political or economic forces."

They go, or are taken, to Pearl Harbor by tour bus, public bus, or rental car, on interstate freeway H-1. The Dwight D. Eisenhower Interstate and Defense Highway is a system of Ka'ai's "longer roads." It was originally built for national defense, to provide rapid ground transportation between U.S. military bases in case of enemy attack. Although that need has now passed, the mandate of the Interstate and Defense Highway system "plays [other] percentages" and continues the military's writing of the land. Locally, the system provides the main, usually only, transportation corridors for all vehicular traffic. Highway construction provides some temporary relief to commuters facing an immense increase in auto traffic, equally temporary jobs for residents, and major contributions to political campaigns.[6] The freeway system (H-1 through H-3) is marked with such exits as Pearl Harbor, Hickam Air Force Base, Kaneohe Marine Corps Base, Fort Shafter, Schofield Barracks, and the *Arizona* Memorial. As indicators of everyday destinations, the signs normalize the ubiquitousness of the military presence on O'ahu.

Visitors who enter via the Pearl Harbor gate drive through navy housing, a pleasant area of one- and two-storied buildings with long sloping roofs and grassy lawns planted with monkeypods and palms. It is an area like the suburbs, only with a difference. It is also public housing, as is the White House, but in somewhat different ways. Common to all of these spatial inscriptions is the face of the nation-state peeking through, creating different public identities while proclaiming its eminence. All routes to the memorial terminate at the large parking lot shared by the USS *Arizona* Memorial Visitors Center and the USS *Bowfin* Submarine Museum. The USS *Missouri* will soon be added to these spectacles.[7] Here at the water's edge, the spatialization code shifts abruptly from the domestic to the martial: buoys, large armored ships, the water of the harbor, cranes, rectangular gray buildings, concrete ramps, small craft, rifles, stenciled acronyms, uniformed men and women, antennae, military vehicles, flags, masts, chain-link fences, warning signs, guards. This is a portrait of the "singular face of state power" (Klein 1989: 103): the national security state in its battle dress, simultaneously suggesting the world as a place of dangerous foreignness and vulnerable boundaries. It is a semiotic summons of the tourist to the reigning narrative of war and victory; it is a pedagogue proclaiming the might and legitimacy of the American security state, both threatening and reassuring; it is the stern father permitting his children to see his gun collection, which he normally keeps locked away and shows only to his friends.

As a significant national memory site, the Pearl Harbor/*Arizona* area is a semiotic thicket. The memorial is a national monument operated by the U.S. Park Service, and also a field charged with desire. There is much at stake at Pearl Harbor for the state, for it must at once assert its self-evident legitimacy as well as claim the ability to maintain its borders intact in the face of tragic evidence of its inability to do so. This is where the tourist gaze, national narratives, myths, identities, and semiotic codes converge to produce a site where seeing is believing.

The first thing visitors see as they approach the visitors center from the parking lots is an anchor and a section of its chain leaning against a wall. Recovered from the *Arizona,* these are the first of the primary icons supporting the memorial's pedagogy. Metonymically evoking the lost ship and men, the anchor appears as an innocent but tragic reminder of the threat and chaos that exist beyond the nation's borders. Semiotically coded as "the real," it anchors the dominant narrative enveloping Pearl Harbor as the site of a cowardly attack by foreignness on American innocence and American soil. The interpretive work necessary to keep the sneak-attack story paramount illustrates the labor required from tourist discourse to avoid seeing the world as a "configuration of political or economic forces" (Culler 1988: 153). Reminders of the appropriations, struggles, and subjugations necessary to the formation of the world as a boutique would scare off the customers as well as undercut the self-evidentness of the nation-state. The theme of innocence denies the history of decades of contention between the United States and Japan over trade, military goals, and foreign policy and, in the same gesture, reaffirms the givenness of the state system. "American soil" obliterates American appropriation of Hawai'i and the subsequent colonial order in Hawai'i at the time of the attack. The subjugation of Pearl Harbor itself began a century ago as the first of the not-at-all obscure objects of American military desire. That desire's trajectory ran from developing Pearl Harbor as a coaling station, then as a simple naval station, then to the dredging and construction of a major dry dock, and, finally, to the developing and fortifying of today's complex fleet and command headquarters.

All points along this arc of desire's path required vigorous pacification of the indigenous inhabitants and Ka'ai's "perforated strip" of the harbor itself. In earlier times Pearl Harbor was a large irregularly shaped body of water indented along its edges by a number of inlets, each of which had specific names. Bearing the name *Pu'uloa* (a long cinder cone or hill), the harbor was the site of forty *loko i'a* or fishponds, part of the 360 estimated to have been in existence in Hawai'i at the time of Cook's arrival. The annual production of fish, an important element in the Hawaiian diet, was estimated to have been about two million pounds (Ali 1991). The waters of *Pu'uloa* were "the stuff of legend," supporting enormous schools of mullet and nehu, rich in oysters, clams, and mussels. In its pacification, what Ka'ai calls the "lengthening of its road," the fish have gone, the fishponds are nearly all filled in, and the area is now "part of the real estate occupied by the navy in connection with its activities at Pearl Harbor" (Tummons 1991a: 4).

If read differently, the anchor could continue to evoke the sunken ship and lost men, but in a way that denaturalizes the national security state, showing how it participates in constituting the kinds of dangers against which it ostensibly defends us. It could expose war to a critique, making it problematic or less desirable, in the word's fullest sense. It might implicate the nation's military and political leaders, questioning the politics of the promotion process within the military.[8] It might make an issue of warfare tacticians, who anchored eight battleships in two tidy rows in such a vulnerable place and at a time when the probability of war seemed very high. But these alternative readings are attenuated by the pedagogy of attacked American innocence, which holds the anchor and chain in place in the discourse of war and the national security state.

Built around a large open courtyard, the visitors center offers the expected tourist amenities of restrooms, drinking fountains, benches, and soda machines and is signed throughout in both English and Japanese. It includes the inevitable bookstore/gift shop, whose merchandise reflects the battlefield tourism market: numerous books on and illustrations of the December 7 attack, World War II, and the Pacific campaign. Other markers of the real include colored slides, flags, models of aircraft and ships, key chains, facsimile newspapers, postcards, sweatshirts, and film.[9] The bookstore merchandise is an active agent of myth, for the history it offers for sale is a particular interpretation posing as History (Barthes 1978: 129). Telling a story that is both frightening (the costs of war) and reassuring (the might of America), myth's sly structural move makes one possible story into the only legitimate one. The bookstore facilitates an easy slide for tourists back and forth between the identities of citizen and consumer; it simultaneously relays the national myths and authorizes the pleasurable entitlements of the scopic regime. Since the real must be marked by signs in order to be apprehended as the real, one part of the tourist quest is for the "experience of signs." Key chains with anchor replicas attached give the anchor its authenticity and signal "I was there" for the visitor (Culler 1988: 159).

Most tourists carry cameras "to make a record of their trip and to show others back home 'that they were really there'" (Little 1991: 156). Both key chains and photos are personal signs anchoring them in the normalizing narrative of war. The scopic regime's technology invites tourists into active participation in the interstices of national narratives; each of the many visitors takes his or her own individual pictures, which look a lot like everyone else's. Recruiting themselves into the available story line, tourists who believe they have framed pictures of the memorial might better understand themselves to be framed by it.

Next to the gift shop is an exhibition gallery paneled in koa (*acacia koa*), a rare wood grown only in Hawai'i. As a "natural" material, it denotes authenticity where plastic or a painted surface cannot. The use of a native wood here domiciles the navy, codes it as a native, not as a part of the colonization of Hawai'i. Koa also symbolizes economic power and prestige and is used for flooring, furniture, paneling, and, occasionally, for outrigger canoes and ukuleles, but not for *most* floors, furniture, or wall panels. Native Hawaiians used koa in making outrigger canoe hulls from trees selected and cut according to prescribed religious rituals (Kamakau 1976: 118–22).

Today, koa forests have become very scarce, in part because the tree is vulnerable to a destructive pest but even more because of private and public land practices.[10] The pacification steps necessary to reproduce the world as a series of shopping opportunities are leading to the replacement of koa forests with koa plantations entailing the predictable loss of biological diversity and self-reproduction. Forests reseed themselves; plantations do not. A variety of plant species, despite different spatial, nutrient, and sunlight requirements, becomes a forest, an order with its own eco-logic. Forests provide cover, attractions, and food for the birds and insects that play roles in pollination and reseeding; the flora of the forest age at different rates; the forest is always in process, its denizens "knit[ting] themselves into the elements of their environment" (Ka'ai 1987). Plantation trees are an age cohort, planted and cultivated to produce the "best"-shaped trees and an easy harvest. Diversity here is disorder; it interferes with the percentages.

In Hawaiian, *koa* can also mean brave and bravery, fearless, valiant, and warrior. The visitor center's exhibition gallery, in which koa has such an active semiotic life, promulgates the themes of heroes and martyrs: paintings and drawings of World War II sea battles, the "rivets and buttons" of navy life, marine landscapes, and portraits of admirals and ships, both afloat and afire. Exhibition cases contain diagrams, models, and photos of the launching and refitting of the USS *Arizona*. Traces of its crew are found in photos of such male-bonding venues as "Battleship Life" (band concerts, athletic competitions), "Hawaiian Liberty" (sailors and girls, hula dancers, grass skirts, beaches, tourists arriving on the *China Clipper* and the *Malolo*), and "Battleship Row" (concerts by the ship's band, a portrait of the *Missouri* with all guns aimed ahead). Recruitment posters from an earlier age timelessly urge young men to join the navy, learn profitable trades, and visit strange lands on "cruising vessels of the U.S. Navy."

Just as no look is innocent, neither is any gallery exhibition simply a random collection of items or solely the mark of a good scavenger. As the eye sees through cultural frames, visitors to a gallery confront a reality framed through the objects exhibited and the relationships between them.[11] Each exhibition metonymically stands in for some element of the Pearl Harbor narrative, producing the exhibit's apparent realism. The memorabilia of Acting Paymaster

Paxton Carter construct the seemingly full story of a hero. His eyeglasses, wallet, high school diploma, dress blues uniform, letters, Order of Neptune, purple heart, and the telegram informing his wife of his death on the *Arizona* are signs describing a martyr's trajectory. Carter is himself an absent sign that myth has grasped tightly, emptying his life of breath, retaining only those signifiers that save and relay the historical narrative. Carter cannot speak; he is represented. About Carter the man we know almost nothing except where and when he died. He is an absent presence on whom has been written a story. The people and conditions that shaped his life—his parents, his boyhood, friends, experiences, education, his decision to enlist, whether his heart lifted at the sight of the ocean, the shrill of a bosun's pipe, or a puff of a Camel cigarette—are unnecessary for the story to work. Myth work makes Carter continue to pull duty; he is a martyr who enlisted "in the service of his country" for which he "gave" his life. Without myth's work, he might have been a young man who, faced with bleak employment prospects in pre-World War II America, joined the navy less to serve his country than to have a job or a place to live. The sailor's personal effects, many of which were issued to him by the navy, come to stand *for* him. The material signs of Carter in the case (his eyeglasses, diploma, medal) are myth's way of keeping him on active duty, endlessly deferring his having had a life in which he lived, laughed, loved.

War and its deaths are a high-stakes legitimacy game for any state. Myth's work is invaluable in transforming the deaths from the torn-flesh register to the ethereal, and, in giving a coherence to war and its prosecutors, solidifying the central givenness of the state and the fixity of eternal danger. When Paxton Carter's wife received the telegram, "We regret to inform you . . ," her loss was recruited into the state's order-producing narrative. "We" gives organization and direction to war, suggesting that it consists of the unified and synchronized movements of an army by war leaders according to a rational, carefully prepared plan. In recent times, the military sends an official spokesperson to break the news. Difficult as this must be for those officers, as well as the relatives, the messengers remain within the legitimating frame of the war. They have no voices of their own. No matter how the state's message of the death of a particular and unique person is delivered, by telegram or a service member, its language can only be monoglossic.[12]

The hero myth skimps death as well as life. Had Carter died at another place and another time—as a sailor or civilian—civil or military authorities would likely have been vigorous in their prosecution of the case. Even had his death been accidental, a serious effort would have been made to locate the responsible agent. But Paxton Carter's death on December 7, 1941, fell immediately into the over-determined grasp of the military's mythic language of death. For the men below decks of the *Arizona,* the question of responsibility was/is immediately referred to the attacking enemy, their grisly deaths enshrouded in heroic phrases: "gave his life," "fell in action," "sacrificed himself," "died gallantly."

A further exhibition at the museum records sailors/heroes decorated for bravery: photos of one officer (a noncommissioned officer at the time), a recipient of the Medal of Honor, are displayed here with accompanying descriptions of his gallantry; next is a photo of an enlisted mess attendant whose courage and action won him the Navy Cross, a decoration of the second highest rank. Myth lurks in the photo captions, which indicate that even mess attendants, who normally do not have combat duties, respond as heroes when their ship and comrades are attacked. Myth's work here covers over the institutional racism of the navy at that time, while it presents the hierarchy of navy command as normal. The Medal of Honor winner was white and the mess attendant African American, the photos reveal; but that racial distinctions were as much a naval practice as button flies and black shoes is unremarked. The gaze is directed to the mess attendant's use of an automatic weapon against enemy planes, not to the naval practice of assigning black men to duties that precluded their contact with and use of guns.[13]

Representations of women in the gallery exhibitions are few. An occasional well-dressed woman christens a ship; more appear as well-dressed background wives behind the front row of important men in uniforms and business suits. Other women wear fewer clothes; they are hula dancers, or women teaching men the words to "Aloha 'Oe," or other young women looking decorative. In their several ways, calling on power's central vectors of class, race, and gender, they serve the state and its warriors even here; the center and memorial are as much about the ship of state as they are about a ship of the fleet, the first a white and masculine vessel, the second white and perpetually feminine, both propelled by the master myths of national preeminence and warrior heroes.[14] The absence of women in the gallery is less an anachronism than a mode of thought, a way in which historical spatial conceptions have either restrained women or eliminated traces of them. Elizabeth Grosz (1995: 121) reads Luce Irigiray's work on space and spatiality to see a

> tendency in phallocentric thought to deny and cover over the debt of life and existence that all subjects, and indeed all theoretical frameworks, owe to the maternal body, their elaborate attempts to foreclose and build over this space with their own (sexually specific) fantasmatic and paranoid projections.

The male world thus constructed, she argues, is "implicated in the systematic and violent erasure of the contributions of women, femininity, and the maternal" (121). Refusing to acknowledge their debt to their origins, men project their own interiors outward, building an "intelligible" universe, producing knowledge, truths, and agents. These "paranoid projections" of "hollowed space" require women as supports to take care of "the private and the interpersonal" (121). The logic underlying this "intelligible" universe is similar to that of Sam Ka'ai's mainland dwellers, who solve problems with "longer roads and percentages" (1987). Anne McClintock's reading of Irigiray similarly suggests that men's refusal to acknowledge their maternal origins, combined with their own lack of such an originating space, leaves them anxious about origins. As compensation, men "diminish women's contribution . . . reducing them to vessels and machines—mere bearers—without creative energy" (1995: 29). McClintock links this anxiety over origins to the acts of baptism and discovery, male rituals that assuage their obsession by arrogating the generative power (of naming) to themselves (29). The projection outward of such an imperial appropriative power is a "spatiality that reflects their own self-representations" (Grosz 1995: 121). From this perspective, the absence of women in the gallery is ontological rather than anachronistic. The intelligibility of the visitors center's manly narrative rides on women's modest domestic presence and looming background absence. The events worthy of chronicle occupy the spaces of martial action while only the ships are female.[15]

Because it has to accommodate one and a half million visitors annually, the visitors center is nearly always crowded. Visitors must take a short boat ride to the memorial for which they receive tickets at the entrance. While awaiting their turn, the crowds may check out the bookstore and examine the gallery exhibitions; cameras and camcorders at the ready, they may also stroll the grounds bordering the harbor where all-weather installations orient their view of the harbor and identify significant sites on the island during the December 7 raid. Or they might line up along walls that are hung with dozens of bronze memorial tablets presented by a variety of such American civic organizations as Rotary Clubs, the Italian-American community of Santa Barbara, the Propellor Club, Police Chiefs' Associations, the Pythian Sisters, Elks Clubs, and the National Security Industrial Association. Prior to the trip to the memorial, they file into an auditorium to see a movie about the raid and the deaths. The sound of a bosun's pipe precedes public announcements, such as those notifying the next group of visitors to proceed to the auditorium.

The twenty-three-minute movie is a complete and moving representation of this major American narrative (White 1998: 710). Where the gallery and bookstore provide discontinuous elements of the narrative, and the memorial is starkly bare except for a few powerful icons, the movie brings all the fragments together and fits the icons into the appropriate order. The gallery and bookstore are secular spaces of some confusion due to the bustle of large crowds and commercial exchange, while the memorial is hallowed ground. The movie is the segue from the profane to the sacred. As Geoffrey White remarks,

> the film acquires a kind of sacred significance as an object that is unique to this site. It can only be seen at the Memorial; it is not for sale as a video; and it is not otherwise promoted or commercialized. When shown at the visitor center, there are no titles or credits of any kind. (712)

The structure of the narrative is simple, short, and continuous, a chronology of events moving from "causes, details, aftermath, and, finally, to the present" (White 1998: 712). The film opens with rolling shots of the names of the dead men cut into a wall of the memorial and then asks how this loss occurred. In a different film shown in earlier years of the museum, the names of the men were whispered, adding to the drama. White identifies five episodes constituting the story of the present film.[16] American political and economic pressures and the concomitant rise of militarism in Japan constitute the historical context. Hawaiian innocence replays the tropical paradise and tranquility of a lazy Sunday morning marking Innocence. Attack and death include footage of the chaos of explosions, death, and devastation of the raid. Recovery reimposes the patriotic order of an America rolling up its sleeves to enlist, unify, and prevail. The moral lesson encourages reflections on death and remembrance, celebrating victory but not before including some scenes of the effect of the frightening, murderous work of war on the troops (White 1998).

Weaving these elements into a finished product represented a lengthy negotiation between Park Service personnel, filmmakers, the Museum Association (which funded it), numerous veterans' groups, and visitors. The result is a film that is less bombastic than its precedessor and, as White remarks, opens more reflective space around war's production and its deaths. Although the movie and gallery suggest the reproductive technology of the world as exhibition, the Park Service, filmmaker, and Museum Association purposely avoided creating a theme park where coolers of beer and a Hawaiian holiday atmosphere would prevail (White 1998). Park Service rangers who introduce the movie may, if they are volunteers or veterans, inject their personal perspectives, but leave little doubt that a somber experience awaits the crowd.

After the movie, visitors file out to board the navy launch for the ten-minute ride. The memorial is a long and narrow white structure, hull-like in shape, built transversely over the long axis of the sunken ship, its middle portion quite open to the air and sky (fig. 14). In profile, the structure dips down in the center to just above the water and soars at both ends. Visitors alight from the launch at one end of the memorial, proceed slowly to the other end, and return to the landing stage as another boatload arrives. They will spend about twenty minutes on the memorial. Despite the work of the lecture and the movie, myth has its work cut out for itself in marshaling fickle, unstable signs into domesticating ranks and order. There is a dangerous surplus of meaning in this treacherous space, for it is a tomb, a hallowed place, the site of an explosion that blew bodies and steel thousands of feet.

At the entrance end of the memorial is the lost ship's bell and at the other are all the dead men's names carved on a wall. The effect is that of a semiotic doppelgänger. As visitors move forward toward the middle of the memorial and the open sunlight, they can see a

Figure 14. Aerial view of the USS Arizona *Memorial. Courtesy Island Heritage Publishing.*

rusted remnant of a large gun turret, some barely submerged supports for another turret, the stump of a mast from which the American flag flies, and, under some conditions of tide and light, the dim outline of the *Arizona*'s coral-encrusted hull. Here the visitors are only several feet above the water, the ship's remnants nearly within reach. This point of the tour is the most dangerous one for the myth and can be the most uncomfortable one for visitors. Some of them see past the signs of the real that the rusted bits of the *Arizona* represent and are more powerfully affected by the present absence of the dead men. Some throw leis into the water and watch them slowly bob away. Yet others are taken up with the signs themselves, photographing them, plying the accompanying park rangers with questions about them.

Moving forward, visitors come to the end opposite the ship's bell. Here in the Shrine Room, on a wall of Vermont marble, the palimpsest of war's merciless violence nearly bleeds through again; on it are carved the names of the perished members of the ship's company.

When each batch of visitors comes to the wall, it hesitates to approach too closely. The wall is set back eight or ten feet behind several low steps; it is flanked by flags and roped off by black velvet rope of the sort used for controlling the movements of customers in banks, airline counters, and post offices. Sometimes leis are draped over the rope. Although some visitors pose for photos in front of the wall, none violates its surface by touching or even approaching closely. Occasionally, the Shrine Room is the scene of such military rituals as reenlistments. On other occasions, the ashes of former crew members are interred in the memorial.[17]

The names threaten to signify war as torn humans, ruptured social fabric, survivors' catastrophic sense of loss. The memorial marshals them into the service of the historic narrative and the structuring of a sentimental memory of the war through framing, distancing, and ordering them. In stark contrast to the Vietnam Veterans Memorial, which leaves its narrative open and challenges war's waste, the frame of the *Arizona* Memorial closes the war narrative, leaving it undisturbed. The never-fully-replaced loss that the death of someone close to us leaves is literally palpable at the Vietnam Wall but is abstract and abstracted by the *Arizona* Memorial. The names are untouchable because the wall is remote, guarded by two flags, several marble steps, and the velvet rope. Rather than merely reaching out, visitors would have to reach up as well as act out to touch most names. The palpability of the names on the Vietnam Veterans Memorial makes possible a different memory and invites the recovery of stories of individual persons because the names are at and below the viewers' level; visitors walk among them. The wall is set down in the ground rather than imposing itself from above, making less likely a single closed narrative about war and loss (see Mack 1995). The chronological ordering of the names at the Vietnam Memorial opens reflective room to multiple histories. Closely sharing space with the names, visitors may have their own lived experiences of the lost one(s), ways of remembering where they were and what they were doing when they heard of the loss. They can reflect on what kind of life was no longer possible. The lack of certainty in searching for a name within a chronological order reproduces some of the chaos of war.

The impersonal nature of modern war and mass death paradoxically erases the chaos with a bizarre form of an egalitarian order on

the Shrine Room wall. All the dead men's names are listed alphabeti-
cally, with their rank listed as though it were a simple descriptor or
part of their personhood, and not a significant navy and political
marker. An alphabetical order of names presents abstract, depoliti-
cized data. Lists, like plans and diagrams, speak of order; here the alphabetical order (signifying no playing of favorites) is one more sign
that assists in the rationalizing of war, making it appear that there
had been a coherent plan, rendering combat orderly. Today, it is increasingly unlikely that any visitor will have a personal knowledge
of one of the lost men. Yet even were a visitor to locate a remembered name on the list of 1,177 outlined in black, there is nothing
else for him or her to do except to leave, since this is a tour of the
memorial and the timetable requires it; the domesticating order dou-
bles back on the visitors. The wall is an image constructed for the
visitors to participate in the experience of reverence (Baudrillard
1981). As the visitors consume the sign, it comes to define them; they
are being reverent at the memorial.

If the Shrine Room wall represents one brisk way of sweeping
signs into a normalized, if regrettable order, still other practices and
sites work differently at masking or, following George Mosse (1990:
7), "displac[ing] the reality of war." Mosse speaks here of the ways
in which the reality of the World War I experience was reshaped as
the Myth of the War Experience through such images as monuments,
cemeteries, and other kinds of war memorials. At Pearl Harbor, both
the location and architecture of the memorial cooperate in producing the myth by taking Nature firmly in their grasp, compelling it to
frame the text in the same way that the koa paneling did in the
gallery. The water of the harbor surrounding the immaculate white
concrete memorial, the blue sky, and the sunlight overhead are all
signs of a tranquillity and a permanence completely at odds with
the sudden and grisly deaths of the sailors. These are the "natural"
signs of resurrection, indicating a place of eternal peace, the peace of
the immutable and universal cycle of life where death is natural yet
abstract.

The Nature that speaks here is a severely regulated one, allowed
to set the stage for memory but not to permit natural processes of
decay on men or ship. The wreckage of the *Arizona* has, over time,
been transformed by marine creatures and processes into a partial
reef where marine organisms live; some sea animals such as coral

builders have significantly stabilized the wreck. The water moves, the good and bad odors of a harbor are present, an occasional oil bubble still surfaces, slowly leaking from the ship's tanks. The bodies of the sailors long ago underwent the sea changes effected by the ocean's tides, currents, and animals. There *is* a cycle of life here in these transformations, but the displacement of the reality of war requires resurrection's story of denial,[18] not a story about food chains. Wrapping Nature daintily in the dress whites of immutability also suppresses the violence of the navy's residence in and near the harbor. The contamination of the waters of the harbor, used by the navy for more than a century, is so extensive and concentrated that in July 1991, Pearl Harbor was nominated by the Environmental Protection Agency for inclusion on its Superfund list. Against a minimum qualifying score of 28.4, a study of six sites in the harbor accumulated a score of 70.82 points.[19]

Situating the memorial so that it spans the midsection of the *Arizona* suggests the same liminal ambiguity as churchyards. The dead are both present and absent, their space marked with signs. The hush surrounding the memorial gently nurses along the idea of the ship as a tomb and fathers reverence for the dead. Signs such as the ship's bell, the rusted fragment of the gun turret, and the bit of the main mast support are essential to the narrative, even though they create a moment that could overpower or undo myth's whole project. Such signs allow the realist code and the myth to work together to unify the narrative in which the literal meaning of the signifiers—rusted hull, ocean water, oil leaks—is but a foil for their intention.[20] The signs' duplicity allows intention to hide in the "imperative, button-holing" quality of myth (Barthes 1978: 124), while the signs appear to register simple facts. At the memorial, intention's shrill summons hides at the very moment it asks visitors to look at the rusted steel as the remnants of "reality." Yet the preserved bits of wreckage are parts of the myth, not independent investigators.[21]

The final semiotic gesture of the memorial is the ambiguous upward sweep of each of its ends, capable of being read as either resurrective or triumphant. As resurrection, it continues the theme of war's redemption and promises an afterlife for the dead men. Triumph is about something else. As Marita Sturken (1991: 120) notes, "a memorial refers to the life or lives sacrificed for a particular set of values. Memorials embody grief, loss and tribute or obligation; in

Figure 15. The USS Arizona *Memorial from the perspective of the tour boat. Photo by Phyllis Turnbull.*

doing so, they serve to frame particular historical narratives." The death memorialized here, just as the name suggests, just as the rusted scraps suggest, just as the American flag flown overhead suggests, is that of the ship (fig. 15). And the ship, in its last suppression of war as bursting lungs and torn bodies, comes in a synecdochic turn, to stand for the 1,177 men.[22] As a metaphor of triumph, the upward sweep of the memorial from the low midpoint signifies victory and brings closure to the tensions of the scene so that war can become a meaningful and sacred event.

Charles Griswold (1986: 689) has written that a memorial is a "species of pedagogy . . . [that] seeks to instruct posterity about the past and, in so doing, necessarily reaches a decision about what is worth recovering." Several elements of that pedagogy have been semiotically unpacked throughout this chapter. The memorial can remember American soil only by a structured amnesia about its provenance.[23] It can "remember Pearl Harbor"[24] only by drawing an opaque screen between memory and the racism that made possible the dropping of firebombs and atomic bombs on Japanese civilians. This pedagogy submerges rather than recovers those dead. For the

American tourists who go to Pearl Harbor "to see where it happened," the memorial's pedagogy anchors American identity in an imagined community whose history is that of a ceaseless, arduous reach for freedom irrationally interrupted by the violence of others.

For those forces that constitute the national security state, the silences as well as vociferations of the site normalize war as an epiphenomenon of the behavior of other states, not as embedded in the ontology of the state itself. It does not ask war to explain itself but accepts war's instigation as something lying within the strangeness beyond our borders, and something to which we must always be alert. In this way, the memorial and Pearl Harbor alibi each other. Pearl Harbor, as the frame through which the memorial is experienced, is simultaneously a sign of America's puissance in the world and a ghostly reminder of its vulnerability, which it must constantly deny. For its part, the memorial is the underwriter for the sneak-attack narrative on which the state's legitimacy rides. War deaths are acceptable only in the language of gallant deaths, heroic sacrifice, and falling in action, not incompetence or waste by leaders. The logic of a sneak attack attributes a widely shared rationality to war; as in chess, war's moves are well known by all parties and are predictable to the degree that all know the game. A sneak attack, however, locates the attacker off the board of rationality, in a liminal zone where human difference has run amok. There is no predicting the moves of a lunatic or fiend. From the outset, the sneak attack was the American narrative of choice (Dower 1986: 35–38). Although today there are no traces of the rabid-dog rhetoric in the movie or at the memorial where the story is played out in the open with rusted metal, ocean water, and fresh air, the deaths are still unthinkable except as a result of a sneak attack. John Dower reminds us how the logic of the "exterminist rhetoric" that spawned the sneak-attack explanation helped to produce the demand for an unconditional surrender rather than a negotiated conclusion (Dower 1986: 37). The construed radical foreignness of the Japanese—sneaky, treacherous, inhuman—absolved the United States of responsibility for the ferocity of battle and the loss of life in the Pacific military campaigns, which saw few prisoners taken.[25] It legitimated nuclear warfare on the people of Nagaski and Hiroshima "in order to save American lives." While the memorial does not itself demonize any national group, it keeps warm the rich broth of American innocence, heroic legends, armed might, and duplicitous Others that makes demonization possible.

As a last element of this semiotic pedagogy, the memorial makes it possible to close the book on war, to weave war into the master narrative about this country's eminence in the world and its ability to imprint its order on others. As Peter Ehrenhaus (1989: 98) writes: "Because acts of closure represent the particular interests of those empowered to commemorate, narrative is selective and distortive; it privileges and advances certain world views over alternative ones." The worldview expressed by Sam Ka'ai, with its gentle invitation to commemorate a different sort of loss, is quite unthinkable at the navy's Pearl Harbor. The narrative closure effected by the *Arizona* Memorial normalizes a particular war as well as war in general by giving war a structure of beginning, middle, and end, leaving "cause and effect intact" (Sturken 1991: 122). The effect is to reserve a place for war and victory in the nation's self-understanding.

6

The Pedagogy of Citizenship

Citizenship in the national security state is located in a militarized triangle of gender, race, and class practices. It is a space of membership and marginality, an uneven space in which some dissension is audible but generally feeble in relation to the hegemonic forces it contests. Dissenting voices are weak in that their traditions are often underthematized, their resources sparse, their languages undeveloped. They are also weak in that dissent itself is confused with weakness, for the dominant story is about strength, might, and inordinate power. Civic virtues such as freedom, equality, and citizenship are threaded into the militarized national narrative of conquest and conversion. Contesting voices are often implicitly feminized voices, doing battle in an arena in which strength has already been marked as masculine, proper, and violent.

The dominant narrative of the national security state, like those narratives of early social contract theorists on the origins of society, has to deny its own beginnings. Before independent men contracting with one another for goods, services, and protection must have come mothers and children, nurturance and dependency, the relationships out of which human beings emerge.[1] Similarly, before the national security state came a variety of political arrangements, none of which contained a necessary telos toward contemporary state forms. Born in violence and the elimination of competing spatial arrangements, the modern national security state reflects its own self-representations

155

and disavows its debts to otherness: to maternal life-giving labor, to kinship practices, to the ways of life made to disappear into the narrative of Progress.

The available spaces of citizenship reflect and perpetuate the dishonored debts to other peoples, other relational arrangements that enable different access to public life, other ways of life with different imaginings and mediations. Relatively little is left to chance in the production of the appropriate citizenry because the citizens are in many ways the resources of the national security state; citizenship practices start with the state's requirements, then ask how people can be fitted or shaped into appropriate vehicles. Our project in this chapter is to tease apart the dominant pedagogical practices by which citizens are created for the national security state, to identify significant relay points for the processing of citizenship activities and requirements, and to open further space for more democratic practices of citizenship to gain purchase. To make a space for such a refiguration requires identification and denaturalization of the prevailing practices of citizenship production, and attention to the social spaces within which such production takes place. Such a "politics of location," in Teresa Ebert's (1991: 135) reflections,

> brings forward a whole host of identifications and associations around concepts of place, placement, displacement; location, dis-location; memberment, dis-memberment; citizenship, alienness; boundaries, barriers, transportations; peripheries, cores, and centers. It is about positionality in geographic, historical, social, economic, educational terms. It is about positionality in society based on class, gender, sexuality, age, income. It is also about relationality and the ways in which one is able to access, mediate or reposition oneself, or pass into other spaces given certain other circumstances.

In Hawai'i's intense public spaces, hegemonic practices of citizenship are actively contested, while they are vigorously prosecuted. The complex displacements and mediations of citizenship in Hawai'i mark the turbulent passings and repositioning of contending political locations.

As the military itself is more concentrated in Hawai'i than in many other parts of the United States, so does a militarized citizenship operate more intensely in various public and private spaces. Hawai'i's senior senator, popular liberal Democratic Senator Daniel K. Inouye, is Hawai'i's most active and visible civic pedagogue of the national security state. Senator Inouye was the first American of Japanese ancestry to be sworn into office in the U.S. Congress. Both his personal narrative and his story of the needs of the state (both the state of Hawai'i and the American national state) are closely, seamlessly intertwined in public discourse in Hawai'i. His official life story blossoms into a political narrative that enables certain shifts in positionality (across barriers of class and race) by relying on stability in others (sex, gender, sexuality); that stands on the marks and memories of dismemberment to forge claims to membership; that transforms alienness into citizenship, moves from the periphery to the center, and amid all this destabilization acts as a sturdy anchor and regenerative force for militarized practices of citizenship. Yet the Senator is not a monolithic figure. His long-standing support for Native Americans, his persistent support for certain aspects of the Hawaiian sovereignty movement, his strong record on civil rights, even his continuous allegiance to a New Deal liberalism that became a dirty word during the Reagan-Bush years—all these mark spaces of some turbulence in the Senator's membership in the hegemonic governing order.

Senator Inouye is Hawai'i's first citizen, Hawai'i's poor boy who made good, Hawai'i's World War II hero, Hawai'i's strongest advocate on the national stage. The Senator is also an important pedagogue, a concentrated space of civic instruction through both example and effect. Surrounding Senator Inouye is an intertwining network of political, economic, military, educational, and social vectors of the dominant narratives of the national security state, bringing together war talk and national security discourse. Daniel Inouye as powerful individual is at the center of many of these vectors of power and influence; Dan Inouye as a story, an icon, is deployed throughout these networks as a representational practice, a vehicle for reproducing narratives and policing their domains. As icon, Inouye emblemizes the pedagogical moments both of war talk's given values of heroism and sacrifice, and of national security discourse's taken-for-granted assumptions about danger, insecurity, and the naturalness of its own view of order. As practicing politician, Senator Inouye operates skillfully in the productive moments of the two discursive terrains, calibrating Hawai'i's political economy and America's national interests around the requirements of technostrategic discourse in the national security state.

CONSTRUCTING CITIZENSHIP

The constitutive linkages between/among citizenship, masculinity, whiteness, and the military are complex, persistent, and never entirely stable. Following a path blazed by John Locke, Mark Kann (1991) traces the evolution of civic virtue in the United States within an evolving triangle of patriarchal fathers, soldier sons, and domesticity. Kann identifies two liberal traditions in America: the classical liberal tradition stressing individualism, equal rights, and property; and the republican tradition of civic virtue.[2] The latter view idealized sober and responsible fathers who wed republican mothers, who in turn raised sons to be soldiers, each in his/her own way sacrificing for the public good. These two traditions require each other: "The liberal ideal was that the influence of self-sacrificing women and exposure to military virtues would transform unruly young males into predictable, productive middle-aged adults who could be trusted to exercise individual rights without fostering anarchy or tyranny in society" (4). Kann sees in Locke extensive efforts to make these two traditions work together, to create men who, lured by the privileges of patriarchal fatherhood, tamed by the ministrations of mothers, and disciplined by the severities of soldiering, could be trusted with extensive individual rights.[3]

Kann's triangulated analysis links older men, younger men, and women in a political economy of civic virtue. Older men are envisioned as sober breadwinners, as heads of families, as owners of property. They are entitled to their rights because their individualism rests on, and is tamed by, women's moral policing and young men's death in battle. They are, in Locke's phrase, "'nursing fathers' who are 'tender and careful of the public weal'" (Locke quoted in Kann: 146). Women, in this symbolic economy, represent domesticity, bourgeois sobriety, control over men's desires. Proper Spartan mothers maintain the home front, agree to be protected, uplift the morale of the troops, honor heroes, mourn the dead, and defer to male authority upon men's return from battle. Young men are the circulating currency in this domestic/civic economy: boys raised by republican parents into appropriate manly fortitude become youth who long for military adventure and are disciplined by military virtues toward a future of productive and trustworthy citizenship. They agree to do the protecting, to keep order or at least be used to keep order, when others outside the legitimate triangle (foreigners, immigrants, Indians, workers, slaves, and women who do not properly join the protection racket) become unruly (Stiehm 1983).

What this has meant in America, Kann writes, is that "the martial ethic in America was an enduring challenge to youth to prove their manhood, practice self-denial, demonstrate obedience, and exhibit the civic virtue that informally qualified them to assume manhood and citizenship in a society that treated masculinity, fatherhood, fraternity, and military service as necessary prerequisites to individualism" (292). For young American men of Japanese ancestry such as Daniel Inouye, the bombing of Pearl Harbor by the Japanese navy massively impugned any sense of their civic virtue. Never securely within the legitimate triangle despite their American citizenship,[4] these young men were not allowed to enlist in the American armed forces.[5] Enlistment, an entitlement to queue up for patriarchal power in the triangulated calculus of virtue, buys admission into the youthful masculinity that is confirmed by "the desire, skill, and duty to bear arms for the state" (Kann 1991: 17). Their way into the order of mature masculinity blocked by suspicions of their patriotism,[6] the young nisei appealed to the other strand of the liberal tradition, that of equal rights and responsibilities, for their right to defend the nation and prove their civic virtue. This appeal worked its way slowly through susceptible levels of the national government and resulted in June 1942 in the formation of the 100th Infantry Battalion, a mixture of nisei National Guardsmen and draftees from Hawai'i. When President Roosevelt declared in 1942 that "Americanism is a matter of the mind and heart; Americanism is not, and never was, a matter of race and ancestry" (Aloha Council 1988a), the War Department began to accept the young men into the army. Calling at first for 2,500 volunteers, the department was inundated by a third (12,500) of the military-aged males of Japanese ancestry in Hawai'i. The call resulted in the formation of the famous 442nd Regimental Combat Team, which, along with the 100th Infantry Battalion, took part in some of the fiercest fighting in Europe: hand-to-hand combat at Anzio, frontal attacks up fortified hills, point-blank shooting into German positions in France. Between them, the two fighting units furnished 60 percent of Hawai'i's fighting forces and 80 percent of its casualties. As Lawrence Fuchs recaptured their achievements, "Their slogan, 'Go for broke,' reflected the spirited determination of men anxious to vindicate themselves and their families" (1961: 306).

The self-understandings that Inouye and his comrades brought to these martial opportunities were more complex than the official story about gratitude and anxiety. At least some of the American men of Japanese ancestry resented the need to demonstrate loyalty. Memory of the emotional sales pitch offered by the military representatives and Japanese community leaders urging Americans of Japanese ancestry to enlist evoked these reflections from the Senator: "Why the hell should I prove my loyalty? But if this is the only way it's going to be done, well we've got to do it. . . . Deep in our hearts we resented this, that we have to prove our loyalty" (Inouye 1975: 5). Inouye further recalled his ambitions to attend medical school, his self-projection of familial responsibility as the eldest son of the eldest son, five times over, and his resentment of the war's intrusion into his professional goals. Young Inouye was categorized by the military as 4C, "enemy alien." Recollecting this experience fifty-five years later, his voice still rose with indignation: "You cannot insult me beyond that" (Inouye 1996). Whatever their personal reservations or resentments, however, there was only one available avenue into the civic arena for Inouye and his cohorts. While the racial dimensions of their journey are obvious, no less central to this story of loyalty and trustworthiness are the gendered practices that underpin it: the "loyalty" that American men of Japanese ancestry were required to prove is a support structure for proper masculinity; to become real Americans, these males had to become real men.

In all, 7,500 men served in the 442nd, of whom 3,600 received medals for battle wounds, 700 were killed, another 700 maimed, and an additional 1,000 seriously wounded (Fuchs 1961: 306). Senator Inouye's bravery while leading a charge on a heavily defended hill in Italy cost him his right arm and earned him the Distinguished Service Cross for valor, the Bronze Star for gallantry, and a battlefield promotion to second lieutenant. The 442nd was ten times cited by the War Department for outstanding accomplishments (Fuchs 1961: 306). In all, these impressive acts of bravery scrambled long-held and defended places and placements, opened paths between alienness and citizenship, altered the exchange rates between membership and dismemberment.

After recuperation from his wounds,[7] Inouye earned a law degree and returned to Hawai'i. Along with others previously marginal to the eligibility criteria of civic virtue, he worked to re-form the Democratic Party in Hawai'i. Still using the equal rights and responsibilities argument of liberalism, the non-Caucasians/veterans/Democrats were able again to reposition themselves, this time from the political periphery of Hawai'i to the center, and to gain access into other spaces.[8] Hawai'i's transition from territory to state in 1959 reordered the meaning of citizenship for the people of Hawai'i, who exchanged access to the American electoral process for subjecthood in the American security state. Inouye's empty sleeve became emblematic of Japanese American and other non-Caucasian entitlements in the local hegemonic order. A short apprenticeship in local politics substantiated his own qualification for national public office and his entrance into the order of the "nursing fathers"—the mature statesmen who tend the "public weal" (Locke quoted in Kann 1991: 146). Senator Inouye's home page describes this movement in a 1962 reminiscence by Representative Leo O'Brien:

> The House was very still. It was about to witness the swearing in, not only of the first Congressman from Hawaii, but the first American of Japanese descent to serve in either House of Congress.
>
> "Raise your right hand and repeat after me," intoned Speaker Rayburn.
>
> The hush deepened as the young Congressman raised not his right hand but his left and he repeated the oath of office.
>
> There was no right hand, Mr. Speaker. It had been lost in combat by that young American soldier in World War II. Who can deny that, at the moment, a ton of prejudice slipped quietly to the floor of the House of Representatives. (Inouye 1996: 2)

Representative O'Brien interprets this dramatic moment as a step toward racial equality, but the transaction is equally open to being read as confirmation of soldiering's importance to citizenship. Inouye's dismemberment was a key to membership in the national political elite, transforming his racial and class alienness into masculine belonging. "He was the type of son any father would be proud of," an aide remarked (Sullam 1978: 22). Race is temporarily suspended, masculinity is achieved, and entry into the arena of patriarchal power is secured.

Inouye's move to his eminence in the Senate, to which he was elected in 1962, was accomplished by "not making big waves"; in not making significant enemies, he became "a seasoned backroom operator capable of dealing amicably with the most diverse factions in the Senate" (Aloha Council 1988a). His national presence includes a series of high-profile performances: keynote speaker at the 1968 Democratic Convention; member of the Senate Watergate Committee in 1973 and 1974; first Chairman of the Senate Select Committee on Intelligence, 1976–78; Senior Counselor to the Kissinger Commission in 1984; Chairman of the Senate Select Committee on Secret Military Assistance to Iran and the Nicaraguan Opposition (Irangate) in 1987. His leadership positions include ranking Democrat on the Committee on Indian Affairs, on the Senate Appropriations Subcommittee on Defense, and on the Senate Commerce, Science and Transportation Subcommittee (Inouye 1996: 2–3). He is the fifth most senior member of the Senate, the third-ranking leader among Senate Democrats, a strong candidate for the top party leadership position, and an oft-mentioned possibility for the vice presidential nomination of his party.

Racial difference has been largely displaced to accommodate the Senator's ascension into predominantly white domains of power. While his consistently strong civil rights record may reflect his experiences as an American of Japanese ancestry, his remarks on foreign investment in Hawai'i suggest a complex mix of racial identification and denial. In a 1990 interview with the editors of the two local newspapers, the Senator criticized laws seemingly aimed at Japanese investors:

> The biggest purchaser of American property is not the Japanese, not yet. They are getting up there. Here in Hawaii, are we going to tell the Canadians, don't buy anything? They don't make a big fuss about it, but they just about own Kihei, don't they?
>
> Maybe they look like us, so we are not fussing, you know? (*Honolulu Star-Bulletin and Advertiser* 1990: B-3)

There is remarkable racial slippage in this statement, which recognizes the underlying racial hostilities in the investment controversy, while the Senator simultaneously positions himself as part of the "us" who look haole. The circumstances under which an Asian American can join the nation's governing class no doubt require such displacements.

Politics Now (1995), an online publication of the National Journal, asks about Inouye, "What does he believe in?" They answer

"The Senate, the Democratic Party, Hawaii, the armed services, American Indians." Speaking of federal policy questions, the Senator described his priorities: "I always try to apply the test: 'Is it in our national interests?'" (interview with Inouye 1996). Senator Inouye's commitments to Hawai'i are reflected in his astonishing record of bringing federal bacon to state kitchens: he has succeeded in keeping nearly all military bases open in Hawai'i when other states face major base closings; he has secured reconstruction of military housing and barracks; he channeled $400 million toward the cleanup of Kaho'olawe, $110 million of which has been appropriated; he has brought home $1 million for a program to prevent the brown tree snake from being accidentally carried to Hawai'i on military planes, $5.3 million for an Air Force optical space tracking station on Maui, $4 million for parking improvements at the Kuakini Medical Center in Honolulu, $2 million for road improvements on the army's Pohakuloa Training Site on the Big Island; his work has provided money for Native Hawaiian educational and health programs, and a $1 million Department of Defense grant to assist displaced sugar mill workers (Pichasie 1996; Omandam 1995a; Dingeman 1996).

There is no end to the list: money for a new shallow-water submarine range on Kaua'i (*Honolulu Star-Bulletin* 1995a: A-10); $16.4 million to form an Army National Guard medevac group on the Big Island; $12 million to continue a health program at Tripler Army Medical Center; $100 million for ship depot maintenance operations at the Pearl Harbor Shipyard; $22.7 million for modernizing the Pearl Harbor Shipyard; $5 million for research and development projects managed by the Center of Excellence for Research of Ocean Studies (*Honolulu Advertiser* 1993, A-2). The Department of Defense 1998 Senate Appropriations bill includes at this writing $35 million for further cleanup of Kaho'olawe, $60 million for the Pacific Missile Range on Kaua'i, and $26.5 million for Tripler Medical Center. The comparable bill in the House includes housing and construction allotments of $70.6 million for Schofield Barracks, $19 million for the Marine Corps Base Hawaii, $17.9 million for Pearl Harbor, and $25 million for the oily waste treatment center at Pearl Harbor (*Honolulu Star-Bulletin* 1997: A-5).

Speaking in 1995 at the East-West Center (EWC), he urged the EWC staff to get over their "hangup against discussing anything having to do with mutual security in the Asian-Pacific region" and admonished them to "stop flooding his office with protests just because a battleship was pictured on the cover of an EWC publication" (Smyser 1995: A-12).⁹ In 1995, he also advised President Clinton that he need not apologize to the Japanese for the A-bombs. "Where will it end?" he asked. "Do we apologize for Iwo Jima, Pelelieu, Guadalcanal?" (quoted in Botticelli 1995: A-1). While visiting Mississippi (where Americans of Japanese ancestry trained in 1943) on the fiftieth anniversary of the end of World War II, he was honored by a flyover of C-130 Hurricane Hunter planes, which he and Sen. Trent Lott saved from deactivation. The unit had been sent to Hawai'i in 1992 to track Hurricane 'Iniki (*Honolulu Star-Bulletin* 1995b, A-8). The Fourth of July was celebrated in Honolulu in 1996 by a big benefit dinner for the Hawaii Foodbank with Sen. Inouye as the honored guest. "Senator Inouye is one of America's finest patriots and has worked tirelessly to improve the quality of life for Hawaii's people. . . . Hawaii cannot honor him enough," said the foodbank's president, Jack Stern. "He has been a true friend—a real Patriot to the foodbank through efforts to obtain federal grants for our capital funding." Military bands greeted guests as they arrived. Individual tickets were $100, but seats at the reserved tables were more expensive: Patriots went for $5,000, George Washington for $3,500, Abraham Lincoln for $2,500, and Patrick Henry for $1,500 (*Foodbank Review* 1996; *Honolulu Advertiser* 1996b: B-3). This partial list of the Senator's accomplishments, spanning merely three years, suggests that Senator Inouye was not bringing home the bacon so much as shipping it, and it reflects his political adroitness, detailed budgetary knowledge, rich institutional memory, and legislative seniority.¹⁰ Equally, it reflects an incredible pervasiveness of military connections and the desperation with which most of Hawai'i's elites and many of its citizens cling to the military pipeline. If Senator Inouye is remarkably skillful at surfing the military appropriations net, we can read his success not only as support for Hawai'i's second-largest industry but also as ominous indication of the extensive militarization of American political culture, of Hawai'i's deep-seated dependence on that culture, and on the resultant impoverishment of spaces for the democratic recalibration of citizenship.

Senator Inouye perhaps unconsciously echoes Locke's expectations of nursing fathers in his own publicly stated self-understanding of his civic office. Paraphrasing Locke, Kann remarks that "political prerogative was akin to parental discretion; it was the exercise of leadership and initiative for the good and only the good of one's subjects. And it was virtually always rewarded by the 'fond' gratitude, if not the open 'love,' of the people" (Kann 1991: 157). Responding to mainland criticism of the $1 million program regarding tree snakes, Inouye cited the danger the snake poses to Hawai'i's environment: "If I hadn't done that, the people of Hawaii should take steps to get me out" (Dingeman 1996: A-1).[11] Inouye's position as Hawai'i's leading citizen calls on and condenses many of the fatherly traits of Locke's nursing fathers, who "encouraged filial affection and gratitude, obedience, and consent among the governed" (Kann 1991: 147). The Senator takes care of Hawai'i. "Inouye Chides Shipyard Workers," proclaimed the *Honolulu Advertiser* in early 1996. Making a personal appearance at Pearl Harbor Shipyard to "soften the blow of impending layoffs," Inouye delivered a "pep talk/scolding" in which he praised the quality of the work, criticized workers for the high costs and inefficiencies of the operation, and counseled civilians losing their jobs due to military downsizing on ways to move into other jobs in the military complex (Kresnak 1996: A-1, A-2). The personal ambitions assumed and encouraged by the liberal individualist tradition are not absent from the Senator's story, but they are notably tempered by the patriarchal obligations of the nursing father. Inouye's own understanding of what it means to be a civic father stresses the particular Japanese flavor of his paternal politics: the importance of honor, of concern for one's obligations, and the mandate "not to bring dishonor for my constituency" (Nader and Fellmeth 1972: 7; Sullam 1978: 22). Reflecting perhaps his esteem for the Senate as much as his responsibilities for Hawai'i, both the man and the icon represent Locke's equation of "founding fathers and political leadership with an elite corps of nursing fathers" (Kann 1991: 155).[12]

The third base of Kann's analytic triangle, the realm of domesticity and republican womanhood, is a less obvious theme in the narrative of soldiering and governing that is the public Inouye. Domestic space by definition is somewhat hidden, rendering public dependencies on personal life arrangements opaque and underthematized. Part of the violence of the national security state is its denial or appropriation of such dependencies, its construal of Independent Men in Charge with no women or children in sight or in discourse. When the young men of the 442nd were claiming entitlement to American masculinity on the battlefields of Europe, their female counterparts in Hawai'i were negotiating the crossings between domestic and public space within the wartime confines of America's citizens of Japanese ancestry. Since Japanese Americans were officially suspect during the war, not because of their conduct but purely on the basis of their Japaneseness, the women of Hawai'i's Japanese community were required to keep a low profile. At the same time, they negotiated the hazards and restrictions of martial law, the fears and resentments of their non-Japanese neighbors, and the prohibitions on Japanese cultural practices, religious ceremonies, and language schools (Okihiro 1991). The job of these women was to make Japaneseness invisible, to keep their obviously Japanese elders from attracting public hostility, and to contribute to the campaign for the enlistment of Americans of Japanese ancestry by erasing any perceptible manifestations of Japanese culture—no kimonos, no Buddhist ceremonies at parks or cemeteries, no tea ceremonies, no public utterances in Japanese.[13]

The sex-ratio imbalance resulting from the huge influx of Caucasian soldiers and sailors aroused parental worries for the safety and vulnerability of their daughters (Allen 1950: 232–33). Under martial law, Japanese women were not permitted into many occupations, but they were allowed to attend the university where many of them majored in education.[14] At the war's end they became teachers while their husbands and sons went to school on the GI Bill. Many women from the local Japanese community contributed domestic war work such as knitting, rolling bandages, visiting hospitals, and aiding in clothing drives (Allen 1950: 350). "When the Army opened the ranks of Women's Corps for voluntary enlistment to women of Hawaii [in October 1944] the Americans of Japanese ancestry constituted 44% of total recruits" (Rademaker 1951: 160). American women of Japanese ancestry may also have found themselves negotiating complex crosscutting loyalties among elders, husbands, and sons. Frequently, the older children in the family were attending universities in Japan in 1942, cut off from the family and subject to the devastating food shortages and labor pressures in wartime Japan. Elders may still have felt strong obligations to their native Japan (Whitney 1995: 46–48, 110–12). Gwenfread Allen observes that "in many households, the father was in an internment camp, the sons in the U.S. Army, and the mother was torn beween two loyalties" (Allen 1950: 352). While the women of Japanese ancestry in Hawai'i were finding their way among these complex racialized antagonisms, they were also marking the boundaries of domestic life that the young Japanese American males needed to leave behind in order to enter the manly world of soldiering.

Mark Kann's analysis of the relations among masculinity, authority, and violence focuses largely on the relations that nursing fathers sustain with soldier sons, spartan mothers, and the voting public, leaving the relations of nursing fathers to one another and to other powerful men unexplored. Tom Dumm's *united states* (1994) brings to the fore these semivisible, underarticulated, necessary, and dangerous relations.[15] He begins by naming American politics as a masculine production, a male order whose achievement requires "an ongoing sexual accomplishment" (52)—the simultaneous production and prohibition of erotic desire among men. Male desire in Kann's account is primarily figured in Lockean terms: for the nursing fathers, desire beckons primarily toward secure property arrangements, legitimate offspring, proper public order, and the good opinion of one's fellows; for young men, desire wears a martial face, a longing for adventure, heroism, and fraternity. Calling on Eve Kosofsky Sedgwick's contrasting homosocial and homosexual realms, Dumm supplements Locke's account with a more sexualized reading of relations among men in power and between men and power. As Sedgwick notes, the two realms are historically produced wrinkles in the continuum of sexuality (in Dumm 1994: 52–53). She calls attention to the cultural practices that accomplish their separation, that keep power relations among men from openly manifesting desire. It takes a great deal of struggle to keep the male bonding of the homosocial world from eliding into the male bonding of the homosexual world. Compulsory heterosexuality and its accompanying homophobia are the keys to policing this border successfully. In the straight male world of public power a man must never love another man. Dumm notes that

> [we] straight men are commonly in positions of superiority over women and gay people, so we feel compelled to construct a homosocial realm that must be homophobic and misogynist in order to allow us to maintain that position. The intense discontinuity between the homosocial and the homosexual is born of that desire for security through superiority. (53)

Working together with other like-positioned men to secure mutually satisfactory power arrangements brings powerful men together in the reciprocal promotion of their interests; valuing these men and the institutions that enable their connivance is key to sustaining the superior position; yet desiring them collapses the fragile house of cards and endangers the security that superiority is meant to guarantee: "Hence the male order in American politics depends on the maintenance of a secrecy about desire among men in power, a secrecy that complicates public discourse about politics and has enabled the associated secrecy of the national security state" (53).

Secrecy follows on the heels of policed desire, responding to the mandate to show a little bit, but not too much, of the erotically charged practices through which men in power get their way. Dumm reaches toward the production of practices that articulate the desires that Kann and Locke attribute to men. Dumm insists that the male body is constantly subject to the reinscriptions of violence producing and reproducing properly masculine men. Real Men have to be able to "take it": learning to drink and smoke cigarettes and calling it pleasurable; learning to deny vulnerability, to endure humiliation and the wet towel snaps of the locker room; learning to sacrifice limbs and life. Referring to his encounters with racism, Inouye remarked, "If you can't take it, you've got no business around here" (Aloha Council 1988a). Real Men who can Take It may well become invested in Taking It, and perhaps also in Dishing It Out—a highly rewarded reciprocal violence unlikely to encourage its participants to rethink its terms.[16] Locke views masculinity as an achievement, well worth the effort; for Dumm, it is a sentence to hard time.

The exchange practices that, following Dumm, lubricate relations between men in power—the combinations of visibility and invisibility, secrecy and confession, loyalty and deference—mark the continued semiotic and material production of the important national security icon, the B-2 bomber. The idea of the bomber began in the late 1970s as a top secret program to construct bombers able "to penetrate undetected deep into Soviet airspace to deliver nuclear payloads" (Council for a Livable World 1995: 1). The B-2 "stealth" bomber sneaked up on the Congress; it was so top secret within the Reagan administration that most members of Congress knew nothing about the B-2 program until 1988, when Congress learned that more than $22 billion had already been spent through the Pentagon's "special access" budget, details of which were classified (Council for a Livable World 1995: 1). Secrecy and exposure work on a variety of levels to produce the public artifact known as the stealth bomber. One level entails the public presentation of the equipment. The citizens of Hawai'i were invited in the spring of 1996 to "the naming ceremony for the B-2 Bomber *Spirit of Hawaii*," which reflected Senator Inouye's steadfast and long-term support for the controversial project. "See it while you've got the chance," proclaimed the three-quarter-page ad by Northrop Grumman in the local evening paper. Senator Inouye and several military commanders were the announced speakers, but the real thrill of the display was the erotic charge of peeking at the powerful and usually closeted technology: "The B-2 Stealth Bomber has a wingspan of 172 feet. Impressive? Just wait until you see it up close. . . . After all, it's not often you get the chance to see the B-2. Then again, that's the way it's supposed to be" (*Honolulu Star-Bulletin* 1996a: A-12).[17]

Here the secret is in plain sight, announced as an invitation to voyeurism, a momentary self-exposure. The military flashes its goods before the public, long enough for the public to know about and be impressed by the sleek and seductive technology, but not long enough to question it. The whole idea behind *stealth* is of course secrecy. But what is the point of a secret weapon that nobody knows about? The public needs to know enough about the weapon to love it,[18] to be reassured by it, to glimpse the magical silver bullet that protects us from a dangerous world. A flirtation is required, not a marriage.

The coy public staging of the B-2 in Hawai'i conceals another level of secrecy and exposure, which lies in the circumstances of the B-2's material production, elimination, and resuscitation. From an initial cost estimate of $21.9 billion for 132 planes in 1981, by 1990 the cost had risen to $70.2 billion. When Congress in 1993 capped the number of B-2s at twenty, the total cost was to be no more than $44.4 billion, or $2.2 billion per airplane. At that unit price, the cost of the B-2, ounce for ounce, exceeds that of gold.[19] A year later, the Senate voted yet more money for the B-2, this time to keep the ability to produce the B-2 in the future. The money was to be allocated to contractors and subcontractors whose work on the aircraft was coming to an end. Senators Bob Dole, Sam Nunn, and Inouye, who worked together to promote the aircraft, defended the funding as a necessary move to "preserve options in an uncertain world" (Council for a Livable World 1995: 3). Senator Inouye also defends the B-2 as a way to "cut the cost of government," in that the B-2 replaces, potentially, the B-52 bomber and the approximately forty-five other craft needed to service and support it (interview with Inouye 1996). After speaking to California aerospace workers during the 1996 campaign, Republican presidential candidate Dole suggested (in an aside to reporters) that he favored producing another twenty bombers. Reacting swiftly if not stealthily, Democratic candidate Clinton agreed to bring the number of the fleet to twenty-one by authorizing the necessary $493 million to convert the first model B-2 to a fully operational aircraft (Federation of American Scientists 1996: 1).

The varied political spaces of the B-2's appearances and disappearances suggest a wealth of wheeling and dealing behind the scenes. Longtime supporters of the stealth bomber were probably not among those congresspersons surprised in 1988 by the news of its production. Presumably immense amounts of negotiating were and are required to navigate the production of a weapon that is both massively expensive and unproven, that was long produced in secrecy, that "melts in the rain, blisters in the sun and chips in the cold" (Scheer 1997: 6), and that has garnered substantial opposition from within the Pentagon as well as from more predictable military critics. While such highly placed opponents of the B-2 as former head of Air Combat Command General John Loh, former Secretary of Defense William Perry, former Deputy Defense Secretary John Deutch, Commander of Air Combat Command Lt. General Richard Hawley, Dr. Paul Kaminski (former Undersecretary of Defense for Acquisition and Technology), and William Myers, formerly with the Congressional Budget Office, advanced reasons for no further funding for the B-2, its builder, Northrop, and its supporters in the Pentagon and Congress managed to continue the dance. "What's the difference between the B-2 and Dracula?" asked House Budget Committee Chairman John Kasich. "Even if you put a stake through the heart of the B-2 it won't die" (Federation of American Scientists 1996: 1, quoting *Washington Post*, September 24, 1995). The configurations of desire that Tom Dumm tracks among Kann's nursing fathers suggest themselves in the exchange practices that must surely lubricate the complex relations among various men in power (Dumm 1994: 74). One hand washes the other. Favors given and favors received. Dishing it out and taking it. You scratch my back and I'll scratch yours. The

stakes in these reciprocal strokings are excitingly high, the maneuvers energetically charged. The thrill of this chase, of powerful men
having their way with one another, is different from the more routine
work of responding to constituent complaints about overdue Social
Security checks or drafting legislation amending a clause in the
bankruptcy law. Not only the homosocial relationships of seasoned
backroom operators, but also the technological props of penetration
and discharge stimulate and seduce. Again, Tom Dumm:

> Being a man in the superior position in this culture means to define
> oneself against queerness, as anyone following our recent national de-
> bate concerning homosexuality and military service knows. But we
> men in the superior position cannot stop sharing in the queerness. It is
> our compulsion, and hence we must purify ourselves to avoid its de-
> bilitating effects, its threat to our power. (1994: 79)

The photo of the B-2 naming dedication in Hawai'i shows U.S.
Senators Inouye and Daniel Akaka, CINCPAC Admiral Joseph
Prueher, Air Force Generals Richard Hawley and John Lorber, and
Northrop Grumman President Kent Kresa lined up in front of the
plane, with the craft's name, *The Spirit of Hawaii,* clearly visible (fig.
16). The photo is a suggestive semiotic confession about proximity
and mutual accord. The presence of two U.S. senators and three U.S.
military commanders, however, encodes the occasion as a devotional
to national security; the presence of the nursing fathers obscures the
openly portrayed links among "special interests," campaign contributions, elected representatives, and Pentagon officials.[20]

Yet another element of secrecy and exposure manifests itself in
the extensive spread of B-2 stealth wealth through the nation (al-
though not in Hawai'i). Citizens whose livelihood depends on the
B-2 include the workers at the final assembly plant at Air Force Plant
42 in Palmdale, California; the flight testing teams at Edwards Air
Force Base in California; the staff of the B-2's operational base at
Whitman Air Force Base, Missouri; the maintenance workers at
Tinker Air Force Base, Oklahoma, Hill Air Force Base in Utah, and
Palmdale; the families dependent on the incomes of the "aircraft
contractor team" at Northrop Grumman's B-2 division in California,
Boeing Military Airplanes Company in Washington, Vought Aircraft
Company in Texas, Hughes Radar Systems Group in California, and
General Electric Aircraft Engine Group in Ohio. "Contracts and

Figure 16. Military-industrial-political leaders at the dedication of the B-2 bomber The Spirit of Hawaii *(left to right): Gen. Richard E. Hawley (commander, Air Combat Command), Sen. Daniel Inouye, Sen. Daniel Akaka, Adm. Joseph Prueher (commander-in-chief of U.S. Pacific Command), Gen. John G. Lorber (commander Pacific air forces), and Kent Kresa (chairman of the board and CEO at Northrop Grumman). Photo by Dennis Oda. Courtesy* Honolulu Star-Bulletin.

subcontractors were spread around to 46 states, in 383 of the nation's 435 congressional districts" (McCartney 1990: B-2). Then there are the miners and refiners who dig and process the ore; the metal workers who assemble the pieces; the scientists and technicians who staff the laboratories, calibrate the instruments, and test the materials; the high-tech workers who produce computer systems; the women and men who type the order forms, keep the books, and clean the offices; the security guards, buildings and grounds workers, and maintenance crews who clean up after the production process; and so forth. Workers and their families in nearly every state depend on the B-2 bomber for their food and rent. This deliberate spread of the B-2 budget among the various congressional districts gives generous supplies of pork to representatives and takes workers hostage. Since the B-2 funding crisis is bound to recur, the bribe of the current contract conceals by only postponing the implicit and inevitable threat of future unemployment.

A final element of secrecy and exposure opens up the hollowness of the national security state itself. The billions spent to produce the B-2—an aircraft that is not as invisible to radar as promised, that sports a radar system that cannot distinguish "between a rain cloud and a mountain" (Scheer 1997: 6), that is unstable in flight, that flies at subsonic speed, and that can be easily photographed from reconnaissance satellites—carry enormous opportunity costs: schools not maintained, medical research not funded, military housing not built, student loans not offered, social services not delivered. Exactly whom is the national security state protecting? The elusiveness of the military threat to which the B-2 supposedly responds and the immediacy of the waste and accompanying social dangers that the B-2 causes expose the fragility of the narrative of the national security state. Just as the beginnings of the modern state lie in unexamined violence against other possibilities of ordering collective life, so does the maintenance of that state reproduce the unquestioned foundational violence. What does not fit into the narrative is, as Dumm suggests, repeatedly repressed to make the story work. The national security state is a narrative about the men and deals that make up its pugnacious dynamics; it is self-referential, articulating and rearticulating the bellicose world that it requires. "The B-2 is essential," claims Senator Inouye. "No other weapon can do the job. With midair refueling, the B-1B and B-52 can fly long ranges, but they can not penetrate heavily defended airspace" (Council for a Livable World 1995: 2).

CAPILLARIES OF CITIZENSHIP

If Daniel Inouye is Hawai'i's First Citizen, his entrance into and success in government imbricated in militarized practices of politics, then the rest of us are in some ways not far behind, for militarized governmentalities saturate American political culture. Following Foucault, we think of governmentalities as spaces of activity, sets of practices that produce, circulate, enforce, and defend The Big Is—the prevailing arrangements of power. The clever pun in government/ mentality condenses Foucault's long-standing argument about power and knowledge: knowledge is power not because one who knows more thus has more power, but because knowledge has its very conditions of possibility in power relations. Knowledge, having no extralinguistic existence, is not discovered. Its production occurs within institutions whose rules governing membership and allocation of resources are not incidental to the other rules that separate knowledge and truth from falsity and superstition. The mentalities that govern in the national security state are those that discipline space into technostrategic discourse, nature into reserves of potential commodities, and people into aggregates of state data. Foucault defines governmentality as

> The ensemble formed by the institutions, procedures, analyses and reflections, the calculations and tactics that allow the exercise of this very specific albeit complex form of power, which has as its target population, as its principle form of knowledge political economy, and as its essential technical means apparatuses of security. (1991b: 103)

In the militarized political spaces of the national security state, citizenship has been reduced to population, policy to narrow forms of political economy, and political action to the practices of security and control. The concept of governmentality enables Foucault's shortcut back to Heidegger's notion of technology as "a positing, ordering, and placing of all beings, here especially human beings as population, at the disposal of an enframing mode of representative-calculative order" (Dillon 1995: 330).[21] Governmentality functions as "a domain of cognition rather than legislation; it is a grid of insistent calculation, experimentation, and evaluation concerned with the conduct of conduct" (Dillon 1995: 330). The image of the grid is central: grids are spatial arrangements that insist, through infinite extension and repetition, on their singular geometric cartography. As a "domain of cognition" governmentality operates not so much by giving orders as by giving birth to order, by making it possible to talk about some things and not others, in some ways but not in other ways. "The conduct of conduct" anchors governmentality not in the direction of a particular action, but in the process of shaping the frame within which conduct can be judged.

Operating within spaces of governmentality, Foucault's apparatuses of security are those dispersed yet connected techniques in which the governing practices of rationality in our society are produced and put into circulation. Apparatuses of security provide the particular detailed methods through which competing or dissenting rationalities are disqualified and dismissed. They materially and semiotically organize conduct within the prevailing terms of the hegemonic governmentalities. They are the methods by which the hegemonic discourses do their work; they organize desire and identity; they micromanage the world of living beings into a warehouse of "standing reserve" (Heidegger 1977: 17); they produce the kinds of people needed by a polity in which citizens are viewed more as passive populations than as active political agents. Apparatuses of security lodge in education, advertising, administration, communication, recreation, and commerce, producing not a seamless web of social control but a complex specification and arrangement of the microconduits necessary to the governmentalities of the national security state.

Effective apparatuses of security require techniques of calculation for the proper ordering of speech and silence, activity and quiescence, memory and habit. Reflecting on the production and circulation of discourse, Foucault inquires:

> Which utterances are destined to disappear without any trace? Which are destined, on the other hand, to enter into human memory through ritual recitation, pedagogy, amusement, festival, publicity? Which are marked down as reusable, and to what ends? Which utterances are put into circulation, and among what groups? Which are repressed and censored? (1991a: 60)

On the national terrain of the security state, and in its concentrated form in Hawai'i, governmentalities of speech and silence, conservation and elimination, reactivation and appropriation are again hidden in plain sight. Sometimes they appear as discrete and identifiable institutions, such as the media, the Chamber of Commerce, the Junior Reserve Officers' Training Corps (JROTC), or the Boy Scouts. Other times they appear in the guise of entertainment, consumption, or festival: sports events, children's toys, parades, holidays, fashions. The broad arena of civil society is so saturated with these practices that effective dissent is visibly limited by their pervasive terms. While these activities of "ritual recitation, pedagogy, amusement, festival, publicity" do not recruit everyone equally, they touch nearly everyone, establishing through familiarity and repetition the meaning of citizenship. Naturalizing bellicosity when possible, supplementing with coercion or bribery when necessary, the spaces of governmentality and their operant apparatuses of security produce and maintain the prevailing memories, anxieties, and desires.

Hawai'i's First Citizen often appears in conjunction with these governmentalizing spaces and requisite apparatuses of security, not as their source or cause, but as a prestigious association, a symbolic representative, a fellow traveler. Reflecting on the relation between particular individuals and the discursive economies in which they reside, Foucault explains that

> there are not on the one hand inert discourses, which are already more than half dead, and on the other hand, an all-powerful subject which manipulates them, overturns them, renews them; but that discoursing subjects form a part of the discursive field—they have their place within it (and their possibilities of displacements) and their function (and their possibilities of functional mutation). Discourse is not a place into which the subjectivity erupts; it is a space of differentiated subject-positions and subject-functions. (1991a: 58)

While Senator Inouye is adept at the backroom discourse of men in power, neither he nor anyone else functions as the controlling hand in militarized discourse and its apparatuses of security. His subject-position is particularly acute, both in regard to Senator Inouye as a politician and to Dan Inouye as a cultural icon, but this acuity largely reflects rather than controls the potent discursive field it inhabits. In some cases, especially regarding those spaces of governmentality that are national (or international) and diffused, there may be no discernible link to the Senator at all. The expressions of militarism that are dispersed through daily practices often do not carry any overt reference to particular men in power, but they enable a "known disposition" of bellicosity (Hobbes 1968: 186), an unease about the world as a dangerous place, an identity wedged in between national pride (We're number one!) and national anxiety (Who will threaten us next?). Some examples of such dispersed daily practices: license plates that carry the emblems "veteran" and "wounded veteran" (in Hawai'i these signs might be associated with our First Veteran); the display of the national flag and the singing of the "Star-Spangled Banner" at athletic events; the organization of the Olympic Games as national contests; presidential profiles on coins; the military uniform on Barbie's mate Ken, and occasionally on Barbie herself; G.I. Joe, and the second and third generations of similar "action figures";[22] toy pistols, rifles, machine guns, rockets, cannons, missiles, bombers, submarines; the private purchase and casual public display of the functioning versions of some of these same weapons; pictures of the president and First Lady in elementary schools, along with the daily pledge of allegiance and patriotic songs; military fatigues and boots as fashion statements; and on, and on, and on.[23]

Our claim is not that Senator Inouye is personally linked to all of these, but that he and they are interactive and reciprocating relay points for the hegemonic national narrative of soldiering and citizenship. Not the individual Dan Inouye but the Inouye effect links the mundane and familiar to that which is heroic, manly, and secure. The Inouye effect "spreads or distends itself over a surface" (Burchell, Gordon, and Miller 1991: ix), in this case the surface of Hawai'i; it naturalizes and makes invisible the different ways in which an activity or art called citizenship has been condensed in a hegemonic discursive field. The Inouye effect is more complex than, say, the Thurmond effect or the Bush effect might be, for Senator Inouye's residency in militarism is not unqualified: he opposed sending marines to Lebanon, invading Grenada, and providing military aid to the Nicaraguan Contras and to the government of El Salvador. A supporter of the Vietnam War during the Johnson years, he later came to oppose it as "racist" (Burris 1988: A-4). Along with the rest of Hawai'i's congressional delegation, Inouye was not sanguine about the Gulf War, feeling that diplomatic efforts should have been more vigorously pursued before going to war (TenBruggencate 1991: A-3). The Inouye effect produces, not an unrestrained enthusiasm for every martial opportunity that comes along, but a more circumspect institutional and discursive marriage of military order with public life. The Inouye effect is in some ways like the Foucault effect, spreading itself over the surface of social spaces, collapsing effect and cause; but politically the Inouye effect is the obverse of the Foucault effect for it makes power invisible rather than visible and disguises rather than questions how things come to be.[24]

We have isolated five different spaces of governmentality that function as potent relay points for the Inouye effect in Hawai'i: the Boy Scouts, the Junior Reserve Officers' Training Corps (JROTC), the Honolulu Chamber of Commerce with special reference to its Military Affairs Committee, the two daily newspapers on O'ahu, and a handful of popular holidays and festivals in Hawai'i. These sites procure for Hawai'i the elements of soldiering and citizenship, manliness and superiority that energize the prevailing discourses of war and security. These active cultural spaces contain order-securing techniques that are both subtle and gross, specifying and arranging individual and collective bodies "in the domain of value and utility" (Foucault 1978: 144). We aim to deconstruct the apparent seamlessness of these militarized spaces and their power/knowledge techniques, to excavate the intense labors that go into producing the appearance of effortlessness and nature, so as to contribute to contesting their hegemony and opening greater spaces for resistances and counterrationalities.

BE PREPARED

It is likely that a majority of American boys do not make their way to adulthood traveling the civic freeway of membership in the Boy Scouts of America, yet the semiotic importance and historical formation of the Scouts as an institution mark a national anxiety about masculinity and a continual need for its social reproduction. From their oath to their laws the Scouts stake out a social topography of militarized masculinity (Seltzer 1992: 149), a geographical imaginary of God/country/homophobia.[25] Small boys in blue uniforms, larger boys in khaki ones do not themselves hew this clearing; rather, they are one of the effects of the turn-of-the-century obsession of such American men as Ernest Thompson Seton "to make a man." Phylogeny's recapitulation of ontogeny was much on the minds of men of that time. Fear of the "progressive effeminacy of men" stimulated a certain hysteria about "the decadent," who was "all that a real man was not: nervous and unsteady instead of restful and strong, exhausted instead of virile, and a libertine to boot" (Mosse 1990: 63). Such men as Seton, Theodore Roosevelt, psychologist G. Stanley Hall, and English Boy Scout organizer Lord Baden-Powell imagined ways for the young male body and character to be sufficiently strengthened and developed to continue human Progress. Seton was concerned to "combat the system that has turned such a large proportion of our robust, manly, self-reliant boyhood into a lot of flat-chested cigarette smokers, with shaky nerves and doubtful vitality" (quoted in Seltzer 1992: 149). Across the Atlantic, Baden-Powell saw the need to correct national deficiencies in such domains as character, health, devotion to empire, and, ultimately, the production of soldiers (Rosenthal 1984: chap. 2). Baden-Powell (Be Prepared) was the more militaristic and ultimately more influential in the American organization than his rival. Seton's less bellicose program of woodcraft assumed that young men needed a knowledge of nature in order to recognize their own human qualities. As might be imagined, Theodore Roosevelt shared the "panic about the natural body in a machine culture" (Seltzer 1992: 152), not departing notably over the years from the observation he offered in 1899 to a men's club that "as it is with the individual, so is it with the nation" (quoted in Seltzer 1992: 149). TR's individual, it might also be imagined, was male.[26] G. Stanley Hall's *Adolescence,* written in 1904, opened the "scientific" study of boys or "boyology" (Seltzer 1992: 152) that was shortly to be relayed through such character factories as the Boy Scouts and the YMCA (with which Daniel Inouye was involved as a boy, since his family could not afford the Boy Scouts).

As a corporate organization concerned with "character reinforcement, citizenship training, and physical fitness," the Boy Scouts of America were incorporated by Congress in 1916.[27] This act brought the blessings of the state on the efforts of numerous institutions producing the arrangements of power, the special techniques and calculations needed to form young men.[28] The social pedagogy ordained by the state to shape this target population is described by David Macleod: "Pervasive age grading [which] had reoriented the issue between adults and boys; instead of a few convulsive struggles for autonomy, there were endless little tests along a finely calibrated course" (1983: 28). One set of these gradations is the vertical distinctions of scout rank: Tenderfoot, Second Class, First Class, Star, Life, and Eagle. At the completion of the requirements for a higher rank, a Scout is individually audited by a board of adult troop leaders who make sure that requirements have been met and suggest ways for a boy to "get more out of Scouting" (Boy Scouts of America 1990: 591). Earning merit badges is another highly regulated, calibrated test; each Scout must get a form for each badge from the Scoutmaster, work with a similarly interested peer, contact the "merit badge counselor" to learn how to "get the most out of the time" spent working on the badge ("A Scout is thrifty"), and submit to the proficiency tests. The remaining six steps in the merit-badge calculus also require adult approval, subjecting the scout to a constant shaping and judging process.

Although determined to shape the strong masculine character among young boys, the patriarchal figures of the organization never had it all their way. Girls, the Charybdis of masculinity, gave them a problem just by their existence. To complicate it in the eyes of some, Boy Scout uniforms of the time included knickers, which made Scouts look like the very thing they were supposed to avoid: sissies. The later introduction of shorts, instead of racheting up the masculinity index, also suggested their wearers were "good boys." Historically, the patriarchs of the Boy Scouts were opposed to the formation of the Girl Scouts (founded in England in 1912) and sought to direct interest toward the Campfire Girls, whose founder, Luther Gulick, charged that it would be "fundamentally evil" for them to copy Boy Scouts. The reproduction of femininity seemed not a problem; girls "were to be more self-abnegating than boys" (Macleod 1983: 50–51). The Boy Scouts of America appealed to white, middle-class parents; immigrant, nonwhite, and semiskilled working-class parents often had different expectations for and needs of their sons. Scout publicity was condescending toward nonwhites, comparing smartly turned-out Scouts to shabbily dressed ethnic minorities (Macleod 1983: 223). Moreover, Scout authorities permitted local councils to decide whether black children could become Scouts, enabling whites to hold veto power in the South (Macleod 1983: 213). The imperviousness of the Boy Scout authorities to race and class differences was evident in their official condemnation of privilege and their belief that a standard, common uniform made the Boy Scouts of America democratic (Macleod 1983: 218).

Scouting plays well in Hawai'i. The principal organization, the Aloha Council, includes most of Hawai'i and extends to American Sāmoa and Micronesia. In 1995, forty thousand persons participated in its activities, including its leaders as well as various young people reached through a nontraditional outreach program.[29] Scout troops are formed through churches, schools, military units, law enforcement agencies, fraternal organizations, and community organizations. An important part of the local program are such community service projects as collecting food for local agencies who distribute it to the hungry, washing post office trucks, cleaning up beaches and roadsides. During World War II Boy Scouts carried messages, helped in aid stations, and acted as coast watchers and security guards (Aloha Council 1995: 9). In the past decade, Girl Scouts have joined them in the important Memorial Day ritual of decorating the graves in Punchbowl with leis. The mostly urban Scouts have a camp in the mountains on Oʻahu, where they have a chance at outdoor camping. Some of their outdoor forays also take them to Schofield Barracks, where they encounter such realities of scouting/martial life as negotiating obstacle courses, crossing rope bridges, rappelling down hills, playing hide-and-seek with camouflaged soldiers, and eating MREs (meals ready to eat). The Scouts make a more formal representation of their accomplishments and abilities at their annual Makahiki Festival.

It is at the local level that the practices of the Boy Scouts of America are effected, where the important symbols of the pedantic regulations and moralistic lessons of the *Scout Handbook* are translated. Some of these technologies of calculation are, in practice, no doubt more observed in the breach. For the boys and perhaps for their leaders, theory and practice may not cohere closely for structural and organizational reasons (see Lum 1990). As volunteers, the Scoutmasters are the ones who have much more than a rhetorical investment in the caring for and teaching of the young. As Macleod observes, the organizational structure of the Boy Scouts of America draws a heavy line between its "aggressive and self-promoting professional bureaucracy" (1983: xi) and its much larger base of volunteer Scoutmasters, who are responsible for putting into operation the nationally standardized program and mission. It is the volunteer leaders who do the useful, sensitive, important work of Scouting: it is they who assist the boys socially, psychically, and physically to solve problems, form relationships, meet challenges.[30] Neither Scouts nor their masters are automatic recruits into the full calibration of Scout/order, but they must find ways to negotiate its specifications. Our point is not that Scouting is a seamless space of indoctrination but, rather, that it is a productive disciplinary space for the organization and articulation of masculinized, calculable subjects.

The annual Aloha Council's Distinguished Citizen Award ceremony clarifies how it is that Boy Scouts are important legitimating symbols, spatial cadets as it were, for the hegemonic interests in Hawaiʻi. Foucault refers to government as a set of relations between "men and things"—"men in their relations, their links, their imbrication with those other things which are wealth, resources, means of subsistence, the territory with its specific qualities, climate, irrigation, fertility, etc." (1991b: 93). The Aloha Council's executive board, the governing board of the Boy Scouts in Hawaiʻi, is a rich space for such links; its eighty-two members include public officials, public relations executives, landed estate trustees, retired military commanders, religious leaders, corporate CEOs, bankers, labor leaders, and businessmen (Aloha Council 1995: 27). It is likely that this concentration of authority and affluence provides an important site for the exchange practices that smooth the surfaces and reduce the frictions among men in power. The Distinguished Citizen Award is presented to a "resident of Hawaii whose leadership in business, industry, government, education, Boy Scouting or community endeavors has contributed significantly to the growth and vitality of Hawaii" (Aloha Council 1988a: 11). This is a position for which women need not apply and no Boy Scout experience is necessary. Senator Inouye was the 1988 Distinguished Citizen, who, in the words of the 1988 *Annual Report,* "personifies the fine ideals of the Boy Scouts of America and the people of Hawaii" (Aloha Council 1988b). Previous Distinguished Citizens have included bankers, businessmen, corporate executives, and elected officials. In addition to the dinner fund's list of donor corporations and their officers, the commemorative program contains the well-known narrative of Senator Inouye's life with its defining way stations: his hardworking and sacrificing parents, his childhood of poverty but dignity, his patriotism and heroism, his commitment to country and community. The Boy Scouts also confer other honors: Silver Antelope, Silver Buffalo, Distinguished Eagle Scout (for which one must have been an Eagle Scout), and Silver Beaver awards.[31] Each of these awards is a marker of approved citizenship, producing and legitimating order-enhancing subject-positions within the governing rationality. At the level of the individual children, the Boy Scouts respond to their social, psychological, and physical needs within an organizational arrangement that promotes disciplinary order and readily militarized self-understandings. At the level of the Distinguished Citizens, the organization reproduces and legitimizes the hegemonic power arrangements of the Men in Charge.

BE ALL THAT YOU CAN BE

Perhaps because of the war in Europe, 1916 was the year that Congress had several thoughts about young American men and the proper spaces for their production. That year, Congress chartered the Boy Scouts of America and also passed the National Defense Act, which established the Junior Reserve Officers' Training Corps. The goal was to "Motivate Young People to be Better Americans" through the introduction of military discipline into public schools (Hawaii Department of Education 1995b: 1). Deleting other possibilities without comment, Congress assumed that "Better Americans" married soldiering to citizenship. During congressional discussion of the ROTC Revitalization Act of 1964, arguments were made to strengthen the motivation by providing for "something" that would implant discipline and respect for order and authority in young men (Lutz and Bartlett 1995: 37, n. 18). Further fine-tuning and expanding the program in 1992, Congress asserted the need "to instill in students in the United States secondary educational institutions the values of citizenship, service to the United States, and personal responsibility and a sense of accomplishment" (JROTC 1995: 1). Congress seeks in these acts a passive definition of citizenship, marked by acceptance of authority and top-down military discipline.

JROTC is open to students (girls became recognized as part of the "Young People of America" in 1973) during their high school years; for some this enfolding into a militarized order could come in their fourteenth year. JROTC's favored apparatuses of security call on transparently militarized techniques. The course of instruction, as devised by the military, covers military drilling and marching, some classroom instruction in other subjects covered in the texts (written and provided by the military), plus optional field trips or activities. Instruction takes place in high school facilities, and students wear their uniforms to school once a week.[32] The curriculum, called Leadership Education and Training (LET), is taught from three levels of textbooks, LET I, II, and III; it may be taken for elective credits in some schools and as core credits in a few. The LET texts cover such subjects as citizenship, communication, American military history, leadership, drug prevention, leadership/lab drill, first aid, map reading, marksmanship, saluting, and physical fitness. Time is most heavily invested in military drill and marching.[33] The LET texts rely on an authoritarian pedagogy—rote learning—that entails reciting a given view rather than assessing facts and making an independent judgment. The content mirrors the form: citizenship is defined in terms of "loyalty as well as respect for constituted authority" (United States Army ROTC 1989: 245). The curriculum, not surprisingly, puts a heavy emphasis on skills and knowledge relevant to the military rather than on the elements of critical thinking—such as formulating alternatives, questioning of sources and information, and making informed choices—required for democratic citizenship.

Various levels of LET texts define the military as the guarantors of citizens' rights and freedoms; therefore, JROTC students are among the most patriotic of Americans (Lutz and Bartlett 1995: part 2). LET teaching materials, which attribute the attainment and retention of American freedom and independence to the military, are notably silent on such questions as how women got the vote. The texts present an uncritical view of U.S. history, which is problematic not only with respect to women but also to ethnic and racial minorities. American culture and history are presented largely as a succession of necessary and successful wars fought by great men for high purposes (United States Army ROTC 1989: 201–31). JROTC history lessons reflect the view that disputes are best settled by force rather than negotiation. Slavery, the internment of Japanese Americans, and the genocidal practices against Native Americans are narrated from the point of view of a necessary central order. The reader is recruited into a history peopled by Us and Others: the privileged "we" secured the order against those sources of insecurity, disciplining and incorporating the dangerous Others. Indians, properly disciplined, can become American Indians; Japanese can become Japanese Americans. The desirable road to citizenship, it is suggested, is soldiering; for minorities it is a necessary path (Lutz and Bartlett 1995: 20–21).

Nationally, only a tiny proportion of JROTC instructors are female, although young women make up as much as 40 percent (or more) of the cadets.[34] Most of the instructors, trained by the military, lack familiarity with a wide array of pedagogical methods; they tend to use the lecture format supplemented with fill-in-the-blank workbooks (Lutz and Bartlett 1995: 14). The compensation of JROTC instructors, who receive active-duty pay, may be higher than that of civilian teachers, who are more likely to have both teaching credentials and experience. School boards often do not closely scrutinize instructors' backgrounds or the curriculum (Lutz and Bartlett 1995: 15). The masculine culture of the military stands at odds with a culture that at least some schools might want to promote: a non-sexist environment and an education of the young that promotes the values of critical thinking and peaceful conflict resolution. JROTC culture and its sponsoring curriculum of efficiency and chain-of-command obedience collide with a democratic education in which citizenship involves independent participation and does not preclude civil disobedience.

Those who promote JROTC in the schools deny that it is a recruiting mechanism for the military and assert that its program has great value for "at-risk" youth.[35] However, recruitment can take a variety of forms. Army Regulation 32 CFR 542.5:3 asserts that JROTC "should create favorable attitudes and impressions toward the Services and toward careers in the Armed Forces" (Lutz and Bartlett 1995: 10). The home page of JROTC at Roosevelt High School in Honolulu reads: "JROTC is a four-year elective program of instruction for the army which emphasizes citizenship, self-discipline, physical fitness, leadership and character development. It provides an orientation to the military as a career option and an opportunity for practical career and vocational education experience." In 1992, Congress increased the number of JROTC units nationwide from sixteen hundred to thirty-five hundred by the 1996–97 academic year. In times of budget cuts, it is of special importance to the military to maintain a high and public-regarding presence among civilians. Significantly, 45 percent of the youth completing the JROTC program enter some branch of the military services, a rate much higher than that of their classmates (Lutz and Bartlett 1995: 10). Among those who join the military, 70 percent do so as enlisted personnel (Lutz and Bartlett 1995: 5).

Unlike regular recruiters who are active-duty military personnel with quotas to fill, JROTC instructors do not need to hustle cadets so actively. JROTC beckons by the lure of its apparatuses of security: the flashy military presence on the campus, the fancy uniforms, the gold braid, the medals, the awards, the flags, the parades with guns. The availability of the program and the camaraderie it offers to many young persons are a powerful advertisement. Colorful JROTC posters are also very enticing. One displayed at JROTC headquarters at Fort Shafter, Hawai'i, has pictures of JROTC cadets with the message: "Show your true colors. Earn a place on the JROTC color guard and you'll make a strong impression at community events and ceremonies. Army JROTC—step up to the challenge." These are not the same kinds of inexpensive posters that hard-pressed teachers cobble together from construction paper and marking pens or crank out on their computers. These are products of expensive advertising firms, funded by the Department of Defense. A local Honolulu newspaper usually covers the JROTC annual spring boot camp with a full-color photo on the front page. Inside pages carry other photos of cadets doing such things as wallowing through mud on an obstacle course and rappelling down a tower. One photo showed them "guided by the Wolfhounds—the 4th battalion, 27th Infantry at Schofield Barracks" (*Honolulu Advertiser* 1995: A-1).[36] With this sort of publicity and these erotic inducements, the hard sell of recruiters is unnecessary.

The JROTC goal of assisting so-called at-risk youth is no more clearly defined than was Congress's original goal of "making better Americans." Judging from the locations of the schools where the growth of JROTC units occurred nationally, *at-risk* is a code for "low-income and/or youth of color" (Rohrer 1994: 36).[37] Despite the claims that the "Army is a good place to start," there is little evidence that military service for minorities is a means of climbing the socioeconomic ladder. Studies of African American veterans show a higher unemployment rate than for their white counterparts, who in turn have a higher rate than white nonvets (Lutz and Bartlett 1995: 7). There is no evidence that the JROTC curriculum for drug prevention differs from or is more effective than that offered in most schools. While JROTC programs offer some good male role models, young persons with drug problems require more than that. Their definition of *at-risk youth* does not include high school dropouts, those with physical disabilities, those with low grades, and certainly not those who are not heterosexual or those confused about their sexual orientation.[38]

The conflation of military training with high school education began in Hawai'i in public schools in 1921 and was part of the curriculum in two principal private schools even earlier (interview with Barrell 1995). In 1994, there were 1,303 cadets in Army JROTC in twelve public schools.[39] Entrance into JROTC requires only that the student be in the ninth grade and in reasonably good health; there are no citizenship requirements. Attracting students to the program is not a problem, according to Lt. Col. Donald Barrell, then state coordinator for JROTC; other cadets (and presumably the military aura) do the job. While the program does not attract many Asian students, Col. Barrell noted that it draws very well from students of Polynesian (especially Sāmoan) and Filipino descent. He further observed that many cadets come from single-parent homes and "less affluent" backgrounds. This is a self-selecting target population of "at-risk" youths, located at the intersection of local race and class hierarchies: these are the ethnic groups in Hawai'i with high rates of unemployment, poverty, ill health, poor housing, and other indicators of social misery. JROTC is accepted as an elective by the State Department of Education. Students' work is graded and many fail; Barrell estimated in 1995 that they lose about 50 percent of the students a year. Stating that "we are not a panacea to the ills of the school," the colonel stressed the importance of his program to the students outside the classroom. The JROTC room "is always open . . . it's a home away from home." The instructors provide male role models and father images during nonschool time and weekend activities such as drills, color guard practice, camps, and leadership training. Col. Barrell emphasized that JROTC teaches students to teach each other, to compete well, and to develop leadership capabilities. "The big thing we do," he added, "[is that] we give the kids some self-esteem" through such visible rewards as medals, ribbons, and recognition. "For a lot of kids no one has ever done anything like that [for them]." Thirty to forty percent of Hawai'i's cadets are young women who are "more mature," better shots, and more likely to take responsibility for necessary paperwork. For Col. Barrell, the end product of four years of JROTC is "a youngster who has a little bit of self-control and self-esteem."[40] Yet as Lutz and Bartlett (1995: 31) conclude, "While many JROTC personnel would clearly like to serve youth, the goal of the Department of Defense of defending its budget, employing its veterans, and garnering new recruits is not consistent with such service." The notion of what it means to "serve youth" is clearly subsumed within a military model and its accompanying techniques of order and knowledge. The thorough integration of soldiering into education elides the possibility that other careers might serve as models of citizenship for high school youth.[41] It is likely that there are a great number of professionals (woodworkers, mimes, musicians, foresters, artists, archaeologists, philosophers, heavy equipment operators, and so forth) who could excite youths, offer alternative visions of competence and responsibility, and enrich the idea of being "better Americans." The Department of Agriculture, for example, is not given an equivalent amount of money to make farmers in the schools, to promote a concept of citizenship based on cultivating land and livestock, or providing people with food. The

National Endowment for the Arts does not have comparable public funds to make artists in the schools or to tie citizenship to creativity, self-expression, or independence of thought.

Youth for Environmental Service persuaded the University of Hawai'i Sea Grant Program Extension Service in 1994 to use some of its federal money to teach good environmental practices to high school students. Three years, ten thousand students, and 120 schools later, there are cleaner beaches, repaired hiking trails, forests rid of invading species, eroded areas rejuvenated in Hawai'i and on the mainland. With the expenditure of very small amounts of money, students have "created healthier communities" (Ambrose 1997: B-1) by learning environmental practices that implicitly tie conservation to citizenship. This successful, if small, example indicates that the military does not fully own the idea of citizenship, yet military institutions have the deepest pockets and military models have the most solid discursive access to sites where the parameters of approved citizenship are produced.

BUSINESS, MEDIA, FESTIVAL

To turn to the 1950 centennial celebration publication of the Honolulu Chamber of Commerce, *Building Honolulu: A Century of Community Service* (Hodge and Ferris 1950), is to turn from the desires of older men for young ones to the desires of older men for themselves. This history recounts its organizational life through "plague and fire, war and depression, and under five forms of government" and concludes that "it's a bundle of unfinished business" (103). Like woman's work, the chamber's work is never done. The difference is that theirs is the producing, circulating, enforcing, and defending of prevailing arrangements of powers. The narrative told by the chamber is an adventure story of efforts to regularize supply and demand on the frontier in the face of epidemics, gold rushes, coinage problems, and dependence on erratic shipping schedules. Its responses to such historically produced challenges and crises can also be seen as scratching out spaces for the production and circulation of hegemonic discourses. What is good for the Chamber of Commerce is good for Hawai'i. When the kingdom's collector general of customs refused to furnish the chamber with import and export data, the chamber appealed to customs commissioners that the data were needed, "not for private purposes, but for the general good of the mercantile community." Shortly afterward, the chamber "took up the routine function of fact-gathering" and became "of service to His Majesty's government for the first time" (3). Shortly after, it sent a petition to the government identifying harbor safety as the "first major community problem" requiring government action. In this way, the Chamber of Commerce became and remains an effective apparatus of discursive security, at times effectively blurring distinctions between private and public decision making, and at other times enunciating standards of appropriateness for speech and silence, memory and habit. In its own words: "It was after annexation that the chamber began to develop into a modern community organization, with a finger in every pie. The chamber's first act after annexation was to place before the Hawaiian Commission its views on government, coasting and tariff laws, needed public works, and other matters" (21).

Other pies included a request for federal public works spending for harbor improvements and lighthouses, support for laying the transpacific cable, and recommendations for a quarantine on incoming animals. The chamber became the trustee of a "public health fund," a self-imposed levy by shippers to clean and maintain the wharves and docks, and it conducted two fund drives, one to support the National Guard of Hawaii. In its own words, the "biggest plum" the chamber pulled from the pies was "the multi-million-dollar dredging project that opened Pearl Harbor to the United States Navy" (26). Still hoping for commercial access, the chamber spearheaded the drive to make the test borings of the coral bar restricting entry into Pearl Harbor. Reversing the contemporary practices that bind defense contractors to Congress, the chamber itself raised $12,000 and borrowed the necessary equipment from the army and navy.[42] Their success sped up by a year the necessary federal appropriations for the dredging and enlarging of the harbor, and the construction of a naval station and large dry dock (27). The completion of Pearl Harbor's initial development was celebrated when the cruiser S.S. *California,* sailing from California, entered the harbor carrying

> more than 250 prominent persons, including Rear Admiral Chauncey Thomas, commander-in-chief of the Pacific fleet; Sanford B. Dole, who was president when Hawaii was a Republic . . . Queen Liliuokalani,

last monarch of Hawaii; E. I. Spalding, president of the Chamber of
Commerce; W. F. Dillingham, President of the Hawaiian Dredging
Company, and Sun Fo, son of Dr. Sun Yat Sen, founder of the Chinese
Republic. (27)

In this glittering array of elites, the chamber is cosily entwined with
Hawai'i's military, political, and economic hegemony. However, the
men of commerce, who found the prospect of a dock close to the
cane fields in the middle of the island very appealing, were ultimately
disappointed when the navy reserved Pearl Harbor for itself. There is
no sign that this disappointment hardened any business hearts
against the military.

Nearly four decades later, the chamber supervised the welcome
of the Matson Flagship *Lurline,* reconverted to a luxury liner after
its wartime use. Its bow encircled by "the biggest lei ever made," the
ship was greeted at sea by navy dive bombers and P-47 fighter
planes, eighty surface craft, a biplane towing a sign reading "Aloha
Lurline," thirteen outrigger canoes, and dockside by a *kahuna*
(Hawaiian priest), the Royal Hawaiian Band, the governor, and the
mayor (59). Not only was the welcome a "symbol of the full-scale
revival of Hawai'i's third industry"(59), but a semiotic choreography of hegemonic discourses proclaiming the display a civic rather
than economic event. Such parades, welcomes, petitions, services,
and information relays are elements of the techniques for organizing
desire and identity.

Today, the supply-and-demand narrative is still very much alive
with the chamber. However, the risks and gains that once were
largely limited to the entrepreneurs setting up shop on Fort Street
and along the wharves now interpellate the whole population,
among whom the costs of the economic fluctuations are socialized
while the profits remain privatized.[43] Still able to distinguish "private
purposes" from "the general good of the mercantile community" and
further persuaded that the latter is indistinguishable from the public
good, the Chamber of Commerce of Hawaii (as it has become since
1967) has no hesitation in speaking for Hawai'i on economic, social,
educational, and military questions. In 1984 the navy announced
that it would accept proposals for home porting a Battleship Surface
Action Group. The chance of such a prize set off a feeding frenzy
among several West Coast cities and Honolulu for the presumed

benefits. The formidable group of persons and interests that scrambled under the chamber umbrella to put together the Honolulu response thought home porting would create two thousand new jobs (Seto 1990b: A-1).[44] Though their work came to naught for Honolulu, the fact that many corporate and political entities could pull together, at such manic speed, to prepare the home-porting document provided an unusually clear portrait of the hegemonic power in Hawai'i. It included the president of the Chamber of Commerce, the governor and lieutenant-governor, the mayor of Honolulu, and all of Hawai'i's congressional delegation. To a man and one woman, they welcomed the opportunity to "serve as host" to the battle group, recited the community's appreciation of the defense establishment's contribution to the state's economy, recalled the era in navy history when it and Hawai'i's kings and queens ruled as partners,[45] reiterated the importance of the strategic location of Pearl Harbor, averred that military preparedness and combat readiness were guarantors of peace, and emphasized that the military was accepted as an integral part of the state. In further supportive testimony the chamber minimized the importance of the nuclear-free movement in Hawai'i, expressed its support for the Kaho'olawe agreement, and championed its "significant role" in obtaining tuition waivers at the University of Hawai'i for the National Guard and the Reserve (Hawaii Economic Development Project 1984: chap. III, part 1). The home-porting effort also provided "the Hawaii business community with a message . . . [that] all branches of the Armed Forces required a single, ongoing Hawaii-based organization to coordinate community response to military requirements and address issues important to any program of expansion of the defense industry in Hawaii" (Chamber of Commerce n.d.).

Thus was born the Military Affairs Council, an unincorporated affiliate of the Chamber of Commerce and a military-industrial combine in its own right.[46] For a number of years the annual budget of the group, whose membership is by invitation, was provided by the State of Hawaii Department of Business and Economic Development, members of the Chamber of Commerce of Hawaii, and businesses related to the defense industry (Chamber of Commerce n.d.). It now operates on an annual allotment from the Chamber of Commerce of Hawaii to serve as the relay "between state and county government, the private sector, and the military community on issues of common interest" such as the cost of living, land use, and public education. The council estimates that the military's economic contributions in 1991 totaled $3.6 billion (Chamber of Commerce 1992: n.p.). Working "on behalf of the defense industry much as the Hawaii Visitors Bureau represents the hospitality industry,"[47] its goals include maximizing the possibilities for local businesses to gain military contracts, encouraging the military to clean up its environmental act, and minimizing the economic impacts of military downsizing (Chamber of Commerce n.d., 1996). It annually hosts Hawaii Military Week honoring "Hawaii's military personnel and their families for their contributions to the State and People of Hawaii" (Chamber of Commerce 1992). Spectacle, amusement, diversion, and publicity are provided by parades, band concerts, open houses at military bases and ships, slide shows, tattoos, pony rides, dinners, luaus, cook-offs, luncheons, weapons fondling, and color guards. Enthusiastically endorsed in the local papers, these events make up some of the discursive and semiotic classrooms that weave together identifications and associations around a concept of place, location, and citizenship (*Honolulu Advertiser* 1994: A-7).

Such practices as these are a variation of the pea and walnut shell game that occludes the presence of the military as a set of problematic social practices by representing it as composed merely of personnel and their families. (In other military contexts families are labeled *dependents,* with all the powerlessness the term implies.) Many military family members may not recognize their lives in Hawai'i in quite the same way as the chamber literature describes them. For younger members, Hawai'i is a hardship post that is expensive, lacking a mainland sense of space, and disconcertingly peopled in racially unfamiliar ways.[48] Many younger personnel are ghettoized, living in on-base housing, shopping at the military PX (post exchange), and using on-base recreational and other facilities (Gray 1972: 15). The average tour of duty for men and officers is three years; for many, a large portion of this time is spent at sea or deployed to Asian countries. This leaves little time for members of the military to make local attachments or develop an understanding of Hawai'i. The more junior their rank, the more difficult military life is, particularly for dependents.[49] For more senior ranks, the situation is somewhat different; it is a difference that local residents may have trouble recognizing as a contribution. The college-age children of military families

pay in-state tuition at the University of Hawai'i. The cost of instruction of military children enrolled in public schools is only fraction-
ally offset by federal school impact money.[50] There are no taxes on
federal property, so military housing does not generate any property-
based revenue for the counties. Nor does shopping in the PXs produce any gross excise revenues for the state. Most active-duty military people maintain their permanent residences in other states, so
they pay no Hawai'i state income taxes. These contradictions rarely
interrupt the constant work of the Military Affairs Council to render
the military's presence in Hawai'i as both natural and benevolent.

The Military Affairs Council is not alone in its supportive labors.
The *Honolulu Advertiser* and the *Honolulu Star-Bulletin,* the principal Honolulu dailies twinned placidly and profitably through a
joint operations agreement, are equally solicitous of the military.[51]
As apparatuses of security, the papers flood the space for the production and reproduction of the dominant discourses of the national state and assist in narrowing "the bandwidth of effective
dissent" (Gordon 1991: 35). Helen Chapin documents this con-
stricting process in recording the often fiery journalistic opposition
to establishment views by numerous English, Hawaiian, and Asian
language newspapers in earlier years in Hawai'i (Chapin 1996).[52]
Her chapter "Battle for Hawai'i's Soul," a careful account of the
underground papers published in Hawai'i as recently as the Vietnam
War, may surprise present-day newspaper readers, particularly when
she writes, "Hawai'i [as] . . . a staging area for the war, generated a
higher ratio of underground papers than the national average, or
about fifteen out of some seventy-five newspapers published during the period" (269). Now there are only two dailies, both owned
by national chains. The *Star-Bulletin* was bought by the Gannett
Corporation in 1971 and sold in 1992 to the Liberty Newspapers
Limited Partnership in order that it might acquire the *Honolulu
Advertiser.*[53] While the two papers may occasionally disagree edito-
rially on local issues such as the importance of the University of
Hawai'i football team or the endorsement of political candidates,
their constructions of the world and the nation are in agreement
about the dangers, tensions, and uncertainties that both threaten
and construct national security.

The newspapers' representations of and news-hole allotments to
the military symbolically naturalize the militarization of Hawai'i.

They are part of the bribe-threat dynamic that produces the "categories of identity and the society of which they are a part" (Doty 1996b: 8). The news hole (the ratio of news to advertising) of the two Honolulu papers is estimated at 20 percent in contrast to the mainland average of 30 percent.[54] Through the eye of this needle pass few results of investigative reporting on any subject. The Honolulu Community-Media Council, after a several months' study in 1991, concluded that both papers "have a long way to go before they become nationally respected dailies of superior quality, matching Honolulu's aspirations to become the economic and cultural hub of the Pacific." Among their recommendations: "Print more news. . . . Increase national and foreign coverage. . . . ban advertising from op-ed page. . . . Make investigative reporting part of daily routine" (Knebel 1991: 1, x). A more recent critic reveals the papers' pathetic dependence for news stories on news releases by public relations spokespersons ("The local press will print anything we send them") and criticizes the traffic back and forth between industries by members of the press and those of the public relations field (Rees 1996: 6). This addictive behavior is ably supported by approximately 115 military public affairs officers (Public Affairs Office 1995: 1–15) and briefings by command-level military officers to businessmen, editors, and government officials.

Press representations of the military are divided between the human face of war talk and the professional discourse of national security. "Navy Wives Endure Their Own Waiting Game," read a headline shortly after the U.S. missile attack on Iraqi targets in early September 1996. A photograph of two women, whose husbands were crewmembers of ships in the carrier battle group in action, occupied about as much space as the story itself. The article sandwiched some factual data about the ships into the narrative of military dependents dealing with their anxiety at such times (Kakesako 1996a: A-4). The bellicose dialect of war talk kept its place in the description of actual military action: "U.S. Jets Kill Iraq Radar as MiGs Answer Raiders" (*Honolulu Star-Bulletin* 1996e: A-1). But the reigning metaphor of the human face is that of the host-guest relationship between Hawai'i and the military. Numbering themselves among the good hosts, the two papers are willing to overlook a great deal of abusive behavior by their guests. Instead of such binge behavior as leaving burning cigarettes on the tables, dropping wet towels on the sofa, pawing the female servants, and urinating in the swimming pool, these guests have contaminated groundwater with diesel fuel, oil, and chemicals in areas on Oʻahu (Tummons 1992c: 4–7), failed to meet clean water and health standards for sewage (Tummons 1991b: 5–6, 8), and seeded valleys with unexploded ordnance (Tummons 1992a: 1, 4–5, 7; 1992b: 1, 7–9). They stored high-level radioactive wastes (Tummons 1993a: 1; 1993b: 1, 6; 1993c: 4–5), cause brush fires, endanger species of flora and fauna, practice open burning of toxic hospital and military waste, and practice open detonation of surplus ordnance (Rohrer 1995: 2–8). In 1990, Hawaiʻi had "five times as many military facilities on the [EPA] hazardous waste docket per square mile as . . . California" (Miller 1990: 2). It is against such a systemic rap sheet that one must read military efforts to mend its ways.[55] At one level, the host-guest metaphor allows an easy slide into the hard-edged national security discourse but in the process stands the host-guest relationship on its head revealing it as a threat; Hawaiʻi is in thrall to the military. Utterances or actions that question the military's occupation of nearly a quarter of the land on Oʻahu (which includes golf courses and other recreational grounds) "send the wrong signal to Washington about Hawaii's attitude to hosting the armed forces" (*Honolulu Star-Bulletin* 1993: A-6). One such "wrong signal," named in the indignant and anxious prose of this editorial, was the effort of Congressman Neil Abercrombie to rein in continued and future military use of Hawaiian ceded lands and to get the air force to relinquish Bellows Air Station.[56] No military planes use Bellows, yet there is strong resistance to converting it to a badly needed general aviation airport on Oʻahu. The marines practice amphibious training at Bellows Beach; behind it is an extensive area of land that Abercrombie has proposed could be used for housing, a great social need on the island for Native Hawaiians, local residents, and military families alike. This modest proposal has earned Congressman Abercrombie the antipathy of both local papers, whose requirements for effusive enthusiasm concerning all things military were unforgivably breached.

Other public sites of hegemonic representations are those found in the plethora of amusements and festivals in Hawaiʻi that marry the economies of tourism with those of war. Beyond the Chamber of Commerce's Hawaii Military Week, other popular holidays and celebrations are marked by the public display of military bodies, military units, military resources. Begun in 1946, the Aloha Festival is "a nine-week, six-Island celebration" (Chapman 1995: 48) featuring a floral parade dotted with military bands and JROTC marching units. A Navy League publication titled *Fore 'n Aft* promotes the annual Hydrofest sponsored by Pearl Harbor and Outrigger Hotels, which offers free boat racing plus games, sports, live bands, and other "family fun."[57] Marching units from the navy, marines, army, National Guard, and air force join with beauty queens and high school marching bands in the annual King Kamehameha Day Parade commemorating the Hawaiian chief who united the Hawaiian Islands through conquest in 1810. Civilian advisory councils, linking military units to local communities, sponsor events such as Military-Civilian Fun Day, featuring volleyball, horseshoes, canoe rides, and other free games and activities. The Hawaiian Open, the major golfing event in Hawai'i, donates some of its proceeds to the Morale, Welfare and Recreation Funds of the four military services. A commercial outfit named Top Gun Tours takes visitors on a whirlwind tour of Pearl Harbor, Hickam Air Force Base, Punchbowl Cemetery, Wheeler Air Force Base, and lunch at Schofield Officers' Mess (Glauberman 1995). Massive V-J Day commemorations marked the fiftieth anniversary of the end of World War II at Punchbowl, Pearl Harbor, and throughout Honolulu: parades, dances, speeches, ceremonies, war narratives, religious services, precision flight demonstrations, receptions, concerts, open houses on bases, weapons displays, dedications.

Senator Daniel Inouye is a frequent guest at these celebrations: he spoke at the tenth annual Memorial Day Eve Candlelight Ceremony at Punchbowl Cemetery in 1991, and at the dedication of the Courtyard of Heroes at Hickam Air Force Base in 1996 (Yoshishige 1991: C-6; Gregory 1995: A-3). At the 1992 Loyalty Day Program in Honolulu the Senator's keynote address proclaimed that "our mothers and our fathers sent more sons and daughters to war than any other territory or state. By doing this we carried out our duty as citizens" (*Honolulu Advertiser* 1992: A-6). It is difficult to find any public celebration in Hawai'i that is not thoroughly intertwined with military markers of "ritual recitation, pedagogy, amusement, festival, publicity" (Foucault 1991a: 60).

The pedagogical practices detailed in the spaces of governmentality from the Scouts through the newspapers to the festivals leave as little as possible to chance in the production of appropriate citizens. As we suggested at the beginning of this chapter, prevailing citizenship practices in the security state begin with the state's requirements and then work overtime to produce the population that fits those requirements. Governmentalities provide "domains of cognition" (Dillon 1995: 330) articulating paths to civic membership that consistently flatten complex social practices into binary orderings and hierarchical evaluations. The hegemonic governmentalities diminish the disorderly contestations entailed in and required by democratic citizenship practices.

Yet hegemony is a work-in-progress, inviting theoretical attention to the possibilities of thinking and acting differently. Stubborn residues of excess and energies of resistance interrupt the "grid of insistent calculation, experimentation and evaluation concerned with the conduct of conduct" (Dillon 1995: 330). To return to our revised version of Toni Morrison's fishbowl metaphor (1990: 17), the bowl, that is, the cultural-political frame that asserts, produces, and disguises militarized self-understandings, is indeed right in front of you, if you know how to look; yet the bowl itself is perpetually unstable and plagued by the leaks that its own efforts at mastery produce. Our methods in this book have tried to offer tools for seeing both the immense power of hegemonic practices (the bowl) and their inevitable incompleteness, their potential vulnerability (the leaks). We have looked to genealogy to help us think against the frame, "to open up what is enclosed, to try to think thoughts that stretch and extend fixed patterns of insistence" (Connolly 1991: 59). Our goal is to make the familiar strange, to make the saturation of Hawai'i's cultural and physical geographies with things military a problem that requires response. We have given broad suggestions about the properties of democratic citizenship—open, ongoing, multiple, contestatory, attentive to its inevitable exclusions—but no institutional formulas, for our task is to help make a space for the possibilities of such citizenship rather than to design it.

The citizenship we envision would nurture the political resources needed to engage in self-reflective understandings of the dimensions of militarism with which this study began. Militarism does its work by producing one kind of order, and mobilizing an array of narrative and institutional devices to naturalize that order while erasing or disparaging other approaches to social coherence. The violences attendant to these productions and erasures are sequestered within stories of Progress and national security, producing an institutional forgetfulness against which individual good intentions can only flail. In some ways the stories of the national security state operate as extortion schemes, elaborate threats, and bribes inducing citizens to accept things as they are. Beyond good intentions, resisting militarism requires citizens who can see differently; who can imagine and sustain a plurality of orders; who can approach otherness with respect and a commitment to negotiate rather than with fear, arrogance, or a determination to conquer or convert. Our vision of citizenship both requires and helps to secure a vigorous public life, one that recognizes the residual violences our preferred orders may contain, while allowing us to struggle toward less violent, more open forms of order. A less violent, more democratic order calls for loosening the demands of microconformity, enlarging the spaces for divergent orientations to find expression, encouraging the relations that might produce respectful strife and acknowledgment of ongoing debts to otherness. Democratic citizenship is also a work-in-progress.

Notes

INTRODUCTION

1. Our use of "hidden in plain sight" follows and gratefully acknowledges Kathleen O. Kane's lead in using it to discuss social practices. See Kane (1994).

2. The use of the diacritical marks the 'ōkina and the kahakō (macron) have political significance in Hawai'i. Hawaiian words require their use. Even though Hawaiian is the second official state language, many publications do not respect the Hawaiian spelling. Our own writing observes the proper spelling in the native language; we also document the accentless spelling used in our many citations.

3. We are indebted to Toni Morrison's acute understanding of the workings of racial metaphors for our own analysis of militarized tropes and conscripted metaphors. See especially chapter 2.

1. TRAFFIC IN TROPICAL BODIES

1. The First Hawaiian Bank estimates the military's landholdings at 5 percent (1993b: 5). No one seems to know the exact figures. Robert C. Schmitt, recently retired state statistician, writes that "still another subject where data are shaky is land ownership, how much acreage is owned or controlled by the federal government, the state, counties, big estates and small private owners" (1995: 7).

2. Again, sources differ. The First Hawaiian Bank (1993b: 10) gives the

following estimation: "About 12 to 15 percent of the state's population is connected directly to the military." Schmitt (1995: 7) says, "National Defense is our second largest 'export' industry and thus worthy of detailed statistical analysis. Regrettably, we lack exact knowledge of either the numbers of military personnel and dependents or total defense expenditures in Hawaii. In 1990 official totals on Armed Forces in the state ranged from 39,936 to 54,001; the same sources reported the number of resident dependents at either 51,727 or 63,215. Defense expenditures were similarly elusive, with 1989 estimates (in billions of dollars) ranging from 2.0 to 2.3 and 2.8 to 3.1, depending on the source."

3. The First Hawaiian Bank (1993a: 1) emphasizes the military's economic importance because, unlike most other economic activity in Hawai'i, the military is not tied to tourism. Reflecting a stunted understanding of economic diversification, the bank declared that "defense is really the only sector of the state economy that now gives us any genuine and significant diversification."

4. We find ourselves stymied by the constraints of the English language. We want to talk about the active coming into being of dynamic places and practices but English anchors us in the seemingly static noun *Hawai'i*.

5. These elisions of sex/gender and race/ethnicity are borrowed from DiPalma (1995).

6. It is likely that the earliest voyagers landed at South Point on the island of Hawai'i. It is the southernmost island and has the tallest mountains, valuable landmarks for those at sea. To date, the earliest habitations have been found in caves in that area. Some archaeologists estimate the age of the first sites as fifteen hundred to two thousand years old. Others make them more recent by a few centuries. See O. A. Bushnell (1993: 3).

7. Close to nineteen hundred species of plants and more than five thousand of insects and small land animals arrived this way (Bushnell 1993: 7).

8. *Olonā*, an important fiber for making cord, grew in Hawai'i before the arrival of the first voyagers. Among the basic elements of their living that they brought were coconuts, taro, breadfruit, *kukui*, sweet potatoes, yams, arrowroot, bananas, sugar cane, *wauke*, mountain apples, *noni*, gourds, and *'awa* (Bushnell 1993: 7).

9. On earlier voyages for the Admiralty, Cook had gone to Tahiti to observe the transit of Venus across the sun in 1769; in a second voyage undertaken in 1772, he had sailed south to Cape Circumcision to discover whether "it is part of that Southern Continent which has so much engaged the attention of Geographers & Former Navigators, or Part of an Island" (Cook 1967, 2: clxvii). On that voyage, he clearly established that there was no continent and no ice-free Arctic Ocean.

10. The marks and inscriptions might include a bottle containing a mes-

sage declaring the land belonging to *Georgius tertius Rex,* or it might be a legend carved on or inset in a tree.

11. Sir Joseph Banks, one of the richest and most powerful men in England, a member of the Royal Society, was an avid, even voracious, botanist who sailed with Cook on his first and third voyages. His plans to sail on the second were thwarted by Cook, who found that the alterations necessary to the ship in order to accommodate Banks's company of fifteen naturalists, draftsmen, painters, and others and their baggage made the ship impossible to handle. The ship was restored to its original condition.

12. Mutiny was, of course, not unknown, but it was a far riskier business in Cook's time than in Magellan's. The disciplining project of the state was well under way when the men under Bligh, who sailed with Cook on his second voyage, mutinied against him knowing the costs they might pay.

13. This is true except of course for the Polynesians, whose ability to navigate by stars, ocean currents, and swells enabled them to sail the Pacific Ocean more than a millennium before Europeans left the sight of land.

14. When Cook groaned after receiving a blow from a club, the chief who hit him "knew that he was a man and not a god, and, that mistake ended, he struck him dead together with four other white men. . . . The bodies of Cook and the four were taken to Kalani'ōpu'u [the chief] . . . and the chief sorrowed over the death of the captain. He dedicated the body of Captain Cook, that is, he offered it as sacrifice to the god with a prayer to grant life to the chief (himself) and to his dominion. Then he stripped the flesh from the bones of Lono. The palms of the hands and the intestines were kept; the remains (pela) were consumed with fire. The bones Kalani'ōpu'u was kind enough to give to the strangers on board the ship, but some were saved by kahunas and worshipped" (Kamakau 1992: 103).

15. Literally, Mauna Loa means the long mountain. "To one unacquainted with the great height of the mountains of Hawaii, this island might appear of comparatively small elevation, for its surface rises gradually from the sea, uniform and unbroken; no abrupt spurs or angular peaks are to be seen" (Wilkes 1845, 4: 114)

16. These statements were among the instructions issued in 1819 to the first company of missionaries to leave for Hawai'i by the American Board of Commissioners for Foreign Missions (ABCFM) (*The Promised Land* 1819: ii, x). Different sets of instructions, with much overlapping text, were used to send off the succeeding waves of missionaries. See Kuykendall 1957: 101–102.

17. The accumulated literature about the missionaries in Hawai'i is extensive and ranges from the self-congratulatory to the severely critical. Full marks for the first kind go to Hiram Bingham, whose first view of the Hawaiians did not change appreciably during his residence: "The appearance of

destitution, degradation, and barbarism, among the chattering, and almost naked savages, whose heads and feet, and much of their sunburnt swarthy skins, were bare, was appalling. Some of our number, with gushing tears, turned away from the spectacle" (Bingham 1981: 81). Another missionary severely critical of Hawaiians and their culture was Sheldon Dibble (1839). More moderate in the sense that he casts his narrative of the missionaries as a "pioneering" folk is MacKinnon Simpson (1993); a classic work of historiography is Ralph S. Kuykendall (1957, 1966, 1967). The limitation of Kuykendall's work is that it is based entirely on English language documents. No Hawaiian language sources were consulted. A recent history of the missionary wives, at once sympathetic and analytical, is by Patricia Grimshaw (1989). Elizabeth Buck (1993) traces the overwhelming political implications of the transformation from oral culture to a literate one by the missionaries. For a critique of the capitalist reworking of Hawai'i in which missionaries and their children were deeply implicated, see Noel Kent's work (1983). Haunani-Kay Trask (1993) writes a series of essays critical of the missionary role in the overthrow of Queen Lili'uokalani and the American colonization of Hawai'i. Her work is also a forceful call for the necessity and right of Hawaiian sovereignty. Lilikalā Kame'eleihiwa (1992) presents a historical analysis of the loss of the land and identifies the missionary-linked persons who were responsible for the loss of Hawaiian land to foreigners after the *Māhele*, the dividing of the land.

18. This border leaked also. While the earliest missionary children were packed off to families in New England, by the 1840s they stayed in Hawai'i and went to Punahou, a school newly founded for them and some chiefs' offspring. Later missionary children were often bilingual but did not display this ability within the missionary circle.

19. One primer used engravings of biblical figures and stories to illustrate the letters of the alphabet. "A is for Adam who ate the fruit Jehovah forbade the two to eat even a little and that is how the first sin came about." Another featured pictures of animals the Hawaiians knew—dogs, pigs, possibly cats, but monkeys as well. Reproduced in Simpson (1993: 38).

20. Sarah Lyman says of the "prodigious" waste of time spent surfing: "You have probably heard that playing on the surf board was a favourite amusement in ancient times. It is too much practised at the present day, and is the source of much iniquity, inasmuch as it leads to intercourse with the sexes without discrimination" (Simpson 1993: 42).

21. One can say that it had become apparent *even* to the missionaries. The Hawaiians knew it earlier; more than four decades elapsed between the arrivals of Cook and the missionaries. The first recorded bodies to make contact with Hawai'i, the foreign ones from Cook's expedition, introduced the first of many microbial bodies to Hawai'i. Venereal disease had spread

from Kaua'i to Hawai'i in the ten months between Cook's first call at the northwest islands and his return to the southeast one from the Pacific Northwest. It was a time, in Foucault's language, of the "effective history" of domination as "the reversal of a relationship of forces, the usurpation of power, the appropriation of a vocabulary turned against those who had once used it" (1977: 154).

With the use of western technology—ships and guns—Kamehameha I conquered the chiefs of the main islands and began the definition of something called Hawai'i. Upon his death, land was not redistributed in the traditional manner; this greatly affected the reciprocal relationship between the commoners and the chiefs. The commoners owed service to the chiefs in exchange for the chiefs showing the proper respect for the gods, without which no one could flourish. The *kapu* system (the system of prohibitions and obligations) had been abandoned by Kamehameha's wife, Ka'ahumanu, and son Liholiho. The traditional way of assuring rightness was not working. Although the Hawaiian people showed considerable shrewdness in dealing with the new ways of commerce being introduced by the early tourists—traders, whalers—they were being silently undone by the germs the first visitors had brought with them. It was into this time of discontinuities, reversals, and so on that the missionaries stepped.

22. O. A. Bushnell writes that the "genocidal decline in birthrates" among pure-blooded Hawaiians continues today. In 1970, they made up 1 percent of the total population; if the decline continues, it is estimated that there will be no full-blooded Hawaiians by 2025 (1993: 271). How many Hawaiians were alive at the time they were "discovered" by the West has political and epidemiological dimensions. Lieut. James King, sailing with Cook, estimated the population to be 500,000, a figure he later reduced to 400,000. Other early visitors, including Capt. William Bligh, put the figure at half or less of King's estimates. In the twentieth century, the figure has consistently been estimated at 250,000 to 300,000, a figure arrived at by what has been called the flimsiest of evidence (Bushnell 1993: 266). More recently, David Stannard (1989), drawing on a range of ethnological, historical, geographical, and demographic data, has radically revised the number upward to 800,000 or 1,000,000. Without contesting the conclusion that the Hawaiians died off far faster than they were reproducing, Andrew Bushnell (1992) has taken issue with Stannard's figures. Citing the published accounts of post-Cook voyagers who reported a healthy population, Andrew Bushnell argues that the kinds of explosive epidemics necessary to Stannard's theory did not take place, except for *ka ma'i 'ōku'u* (literally, "the squatting disease," an unspecific intestinal infection), until the 1820s (1992: 48).

23. To the end, Rev. Sereno Bishop's vision of Hawaiians was uncomplicated by ambiguity. In describing the funeral procession of King Kalākaua, which included members of some Hawaiian societies, he singled out those of *Hale Nauā* as looking "very heathenish in looks and howling," in apparent reference to the high priest of the society who carried objects sacred to Hawaiians (Daws 1968: 264).

24. Within the *ahupuaʻa*, the commoners had several *loʻi*, taro-growing patches, not necessarily side by side. They also had gathering rights within the *ahupuaʻa* that could take them to the uplands for certain plants and into the sea for some kinds of fish. The gathering rights are being reestablished in Hawaiʻi state law, particularly in *Kalipi v. Hawaiian Trust Co. Ltd.* in 1982, and *Public Access Shoreline Hawaii* [PASH] *v. Hawaii County Planning Commission* in 1995. See Rees (1995) and Boylan (1996).

25. *Kalo*, which is also called taro, was the primary starch of the Hawaiian's diet. It is an edible root that after cooking can be pounded into poi. The *loʻi* is an irrigated bed for the growing of taro. For further detail, see chapter 4.

26. The Masters and Servants Law, casually enacted in 1850, included a provision for the apprenticeship of minors. More importantly, it made labor contracts signed in foreign countries binding in Hawaiʻi. Contract workers fleeing the often harsh treatment on some plantations could be imprisoned for the remaining length of their contracts unless they agreed to resume work (Kuykendall 1957: 330). For a parallel analysis of the Masters and Servants Ordinance in Africa and the colonial dilemma of "getting the natives to work," see Doty (1996b: 51–72).

27. The planters sought Cyclops far and near: the Gilbert Islands, the New Hebrides, Portugal, Japan, Norway, and Russia. They found "by costly experience that the only satisfactory and permanent plantation field labor is Asiatic. Caucasians are constitution[ally] and temperamentally unfitted for labor in a tropical or semi-tropical climate. They cannot endure the rigors of arduous physical exertion in the tropics and will not stay on the plantations" (Hawaiian Sugar Planters' Association 1921: 15).

28. The Japanese laborers constituted 45 percent of the field labor force; more ominously, for the planters' fear of engulfment, the Japanese, through attempts to organize plantation labor, "ceased to appreciate the opportunities given them as individuals, and through cohesion as a race, seek, collectively to revolutionize the control of the agricultural industries of the islands" (Hawaiian Sugar Planters' Association 1921: 20).

29. Not all missionaries had the same attitude toward labor: "Some of the members of the disbanded Sandwich Islands Mission and some of their sons went into the sugar business, and they could see nothing much wrong with a contract that contained penal clauses, but other members of the old

missionary family who did not have a vested interest in sugar began to talk about coolie labor and slave labor in the same breath" (Daws 1968: 181).

30. "Mr. J. N. Wright said that . . . in the matter of trashing cane a South Sea Island woman was equal to a Portuguese and only one-third of the expense. As mill laborers they are good, but as cattle drivers they are cruel and inhuman. . . . He would like to try Hindoos. A Portuguese is intelligent but he is cruel. The New Hebrides people seemed to be docile, quick, and a desirable class of laborers. The Chinese were good cultivators and mill hands, but no good as teamsters. On the whole he considered them the most desirable class of laborers.

"Mr. W. H. Bailey said he would like to try Negro labor, but from what he learned when recently passing through Kansas, he felt that it would be impossible to get them without their being accompanied by large families" (*The Planters' Monthly* 1882: 11). "It is not probable that any intelligent planter would prefer to select his laborers all from one nationality, even if a choice were given him. It is too well-known that different characteristics best fit the laborer for different positions. With an allowance for exceptional cases, it is generally conceded that the regular, plodding manner of the Chinaman, and his aptness for imitating or learning from example, peculiarly fits him for the boiling-house, or the routine of hand field work and irrigation. . . . While the Portuguese seem admirably adapted for 'helpers' or assistants in the many mechanical industries pursued upon the plantation, they do not, as a rule, take kindly to the monotonous hand work in the field, nor succeed in the handling of teams. Heretofore we have relied almost exclusively upon the native Hawaiian, who with his natural fondness for the horse and aptitude in handling our half broken cattle, adds the favorable peculiarity of a willingness to work out ukupau [until the work is done] without regard to regularity of hours or time of day" (*The Planters' Monthly* 1882: 29–30).

31. Not without protest from students and faculty, the social sciences building at the University of Hawai'i was named for Dr. Porteus. A thorough critique of the racism embedded in Porteus's work is found in "Testimony of Professor Robert S. Cahill, Department of Political Science, to the Board of Regents, University of Hawaii, April 23, 1975." Twenty-three years later the Board of Regents caught up with Cahill and voted to remove Porteus's name from the building.

32. In 1920 the U.S. Bureau of Education's survey in Hawai'i urged action to deal with the "serious obstacle in the way of the work of the public school, in its task of Americanization—the system of foreign-language schools which exists nowhere else in the United States"(1920: 5). To Americanize the children of Asian ancestry would entail making them "stable, self-supporting worthy members of the society"(35). The report recommended

preparing children in "handwork, manual work, cooking, simple sewing, the making of beds, the care of the house, the making of school and home gardens, the organizing of pig clubs and poultry clubs, and in the use of tools through making simple repairs and through making articles for use in the home" (35).

The reforms led by Governor John Burns and others in the 1960s aimed to change public education in Hawai'i, particularly the university, so that it would elevate children out of, rather than prepare them for, the plantation economy.

33. A British warship commanded by Lord George Paulet forced the Hawaiian monarchy to cede Hawai'i to Great Britain in 1843. While the British, as well as French and Americans, were "interested in" (read desirous of) Hawai'i, the British disavowed Paulet's appropriation and restored Hawaiian sovereignty.

34. As Kuykendall remarks, "the value of Pearl Harbor was well known before these officers made their investigation" (Kuykendall 1966: 296, n. 20).

35. Six men were Hawaiian citizens by birth (but not ethnically Hawaiian) or naturalization. Five were Americans, and two others were English and German, respectively.

36. President Grover Cleveland, informed by the report of a special commissioner he had sent to investigate the overthrow, was opposed to the proposed Senate treaty of annexation of Hawai'i and disavowed the American intervention. Unfortunately for the queen, Cleveland was replaced by William McKinley in whom the annexationists had a strong friend. Although supporters were never able to muster the votes for annexation by treaty, as would seem required by the Constitution, de facto annexation was accomplished by a joint resolution of both houses.

37. The First Hawaiian Bank's idea of diversification is to succeed in getting the U.S.S. *Missouri* installed adjacent to the *Arizona* Memorial at Pearl Harbor (1995: 3). For a more robust vision of diversification, see Rohter 1992.

38. The result, the Hawaiian Hotel, was built for $116,000 near the center of downtown Honolulu, not in Waikīkī. At the corner of Hotel and Richards Streets, it has had an even longer and more useful life than that of which Kuykendall knew. It was converted at the time of World War I into the Army and Navy YMCA; most recently, it was converted into a rather swank office building by developer Chris Hemmeter. Subsequently, it was used during the recent, expensive renovation of the state capitol to house the legislature, governor, and lieutenant governor.

39. Popular photos of the 1920s and 1930s often included American and European notables, male and female and always white, in canoes being paddled by Hawaiian men.

40. These photos were taken by Christian Hedemann, a Danish mechanical engineer who emigrated to Hawai'i in 1878 to work at a remote Hana, Maui, plantation. The photos are excellent given the technology of the time; prints from the black-and-white glass plate negatives make starkly evident the gender, class, and race divisions that supported sugar production. After leaving Hana for Honolulu, Hedemann and his camera recorded, to great aesthetic effect, the mechanical side of sugar production.

41. See also the *Hawaiian Phrase Book* (1968), a compendium of useful phrases for training servants and collecting debts. No author is given for this publication; it could well be attributed simply to The Hegemony. The 1968 edition may be a modern reprint of a book with a similar title published by J. H. Soper, one-time Marshall of the Kingdom.

42. In precolonial Hawai'i, *ali'i* women (of the ruling class of chiefs) were awarded high status and considerable authority, a violation of proper gender arrangements in western eyes. The class and gender *kapus* (rules derived from religion that guided much daily life) that governed this society were focused primarily on food, not sexual intercourse—the mouth more than the vagina was the orifice that mattered. Our thanks to Noenoe Silva for this understanding.

43. We have borrowed this very useful phrase from B. H. Farrell (1982) for use here. We thank him.

2. LOOKING IN THE MIRROR AT FORT DERUSSY

1. While Hawai'i has two senators, most people here would recognize Daniel Inouye as "the Senator." Our use of the uppercase reflects both our respect for his accomplishments and our disquiet over the militarized icon he has become.

2. This and other quotations are taken from display captions and the taped oral narrative at Fort DeRussy Army Museum. Some exhibits have changed since we began this research in 1991. The first gallery, aptly called the Changing Gallery, has changed frequently; other displays have sometimes been revised. Our references are accurate as of March 31, 1998. No catalog is available. Our thanks to Judy Bowman for her help in charting the changes in the museum over time (interview with Bowman 1997).

3. The photo archives at Fort DeRussy contain dozens of photos of Shirley Temple visiting Hawai'i. On her numerous visits the military, according to museum curator Judy Bowman, really "laid out the red carpet" for her (interview with Bowman 1997). Her smiling face and ringlets pair up Daddy's Little Girl with high-ranking military officials and military places, suggesting an all-American innocence safely protected.

While Shirley Temple was not the only celebrity to visit Hawai'i, she was the only one whose presence is so forcefully recorded in the photo archives. It may well be the case that she served as a public relations antidote to the bad press that island tourism received as a result of the Massie case (described later in this chapter). Stories of the alleged assault on a white woman by dark-skinned men likely triggered fantasies that could be negated by the reassurances implicit in photos of feminine whiteness surrounded by uniformed white officers in charge. The Massie case might have had the same effect on tourism in the 1930s that an oil-soaked beach would have on that industry in the 1990s.

4. See also *Women on the Outside,* featured on the P.O.V. series, PBS, July 1996.

5. Since long lines were routine for everything, occasionally even women unknowingly attached themselves to the infamous lines. See Rodriggs 1991: 22.

6. The museum's narrative of loyal citizens enthusiastically buying war bonds elides at least one contrasting story: "War Patriotism tended to ignore the old antagonisms. That was not true of Nanakuli where the homesteaders had developed open hostility toward government of any kind. In September of 1943, volunteer worker Frederick C. Hart, a retired Navy chief living in Maili, encountered this hostility when he canvassed the homestead area selling war bonds. The people asked him why they should buy war bonds to support a government that had never shown any interest in helping them" (McGrath, Brewer, and Krauss 1973: 136).

7. This account is based on Daws 1968: 319–27; Judd 1971: 126–204; Wright 1966. Daws notes that, unlike previous episodes of interracial violence, the Massie story could not be put out of sight (Daws 1968: 319). Judd was governor of the territory at the time when Hawai'i was "in the grip of the Great Depression . . . [and soon] to become the center of a crisis that very nearly saw self-government in Hawaii abolished and replaced by a commission form of government appointed by the President" (Judd 1971: 166). Theon Wright's narrative is based on Honolulu Police files, a report submitted to Governor Judd by the Pinkerton Detective Agency, County Attorney files, trial transcripts, congressional documents, letters, and other documents. This collection and narrative corroborate the concept of justice that allowed four mainland Caucasians to go free after killing a Hawaiian male.

8. Although Governor Judd describes their leave-taking as being "spirited aboard," Wright suggests that a Mack Sennett comedy more appropriately fit the navy's sneaking Thalia Massie aboard the liner from the offshore side. A police officer was stationed at the gangplank to serve her with a subpoena to appear as a witness in the retrial of the surviving four men

charged in the original assault case. The policeman was tackled by a navy captain in the passageway outside Massie's stateroom; with the captain's arms around his legs, the policeman struggled to touch Massie with the papers. Cooler heads soon prevailed; although she was served, Massie sailed five minutes later (1966: 277–78). The case against the enlisted men was finally dropped in 1933 (300).

9. The practice of requiring Japanese Americans who worked in restricted areas to wear special badges with black borders is a specific example of sovereignty with a vengeance. See Shapiro 1993: 2–3. Louise Kubo called our attention to the discussion of this practice in Okihiro (1991: 227).

10. The public circulation of the anticolonial narrative was expanded at the time of the centennial of the overthrow by an extensive article in a local weekly newspaper. See Ferrar and Steele 1993. Also during the 1993 centennial, Aldyth Morris's play "Lili'uokalani" had an extended run at the Manoa Valley Theater, further increasing public awareness of contesting Native Hawaiian accounts of the overthrow. Since these events it has become more difficult for residents of Hawai'i to not know about the overthrow, regardless of the museum's silences. Unable any longer to ignore the overthrow, at least one defender for the Committee of Safety has tried, awkwardly, to defend it. See Twigg-Smith 1997. (Thurston Twigg-Smith is the grandson of Lorrin Thurston, a leader of the Committee of Safety.) Twigg-Smith was involved in a public debate with University of Hawai'i Ethnic Studies Professor Davianna Pōmaika'i McGregor (1997).

11. In fact, annexation was not accomplished by treaty. Unable to secure the votes for a treaty, its proponents settled on a joint resolution.

12. For a detailed account of the machinations of the Committee of Safety, see the report from James H. Blount, Special Commissioner of the United States, to President Cleveland (1893).

13. John Dower (1986) suggests that this is not the way most wars end (37).

14. The museum's audiotape is slightly more direct. It makes brief mention of "cultural influences" on Hawai'i of "increasing numbers of European and American visitors," including "whalers, missionaries, merchants— each leaving their cultural stamp on the island." Suggesting that the "cultural influences" were freely sought rather than imposed, the tape's narrative notes that the "chiefs themselves travelled to Europe and America and acquired tastes for the finer things in Western life," including ceremonial uniforms. The narrative draws a direct and innocent line between the greed of some chiefs and the "influences" of western economic and religious interests. Omitted from this tale are the voices of other Hawaiians, including the thousands of commoners who petitioned their leaders to keep Hawaiian land and governance in Hawaiian hands during the Commoners' Petition

Movement of the 1840s, prior to the *Māhele*. See Kameʻeleihiwa 1992: 329–40; and Kealoha-Scullion 1995.

15. Today, the 25th Infantry Division (Light) is stationed at Schofield. It is the manned combat element for the U.S. Army Pacific. It trains with and in thirty-five different countries throughout the Pacific Basin.

16. Depopulation of many Pacific islands was by then an old story. Removal agents included diseases introduced by early voyagers, and warfare and starvation during World War II. But the removal of the entire population of an island for "strategic" purposes had far more severe social consequences. See Ishtar 1994.

17. As Henry Kissinger was to remark later about all of Micronesia, "There are only 90,000 people out there [in Micronesia]. Who gives a damn?" (quoted in Wilson 1995: 31).

18. Robert Kiste concludes that other factors were perhaps more influential in bringing them to this conclusion. They were used to an exogenous authority structure from colonial experience and a paramount chief; the defeat of Japan had stamped an awareness of the technological might of the United States, which might have meant to them that they had no alternative. It is also doubtful that they ever understood that they might not be returning to their ancestral lands (Kiste 1974).

19. While the Bikinians cannot live on their island, tourists can visit it. A new Operation Crossroads has commenced, this one marrying war to tourism: sports divers can swim among the two dozen sunken ships deposited there by the atomic tests. Visitors live for a week on Bikini, with strict rules against fishing or eating the island's coconuts or bananas. All food and drink are flown in. The emblem of this new Operation Crossroads is a palm tree waving gaily in front of a looming mushroom cloud (Bone 1996: F-1, 2). Reversing the medical/military gaze, the Bikinians now monitor the effects of radiation on wealthy western visitors. One Bikinian mused that "we're very curious about the effects on those people" (Kristof 1997: A-6). While higher than normal, the radioactive levels to which the tourists are exposed are not considered dangerous (Wright 1996b: B-1).

20. Several "safe" methods of cleaning the Bikinian soil of the radioactivity, now taken up by normally edible plant products, have been proposed. The first entails scraping off the top fifteen inches of highly radioactive soil and replacing it. Where to put the hot soil is, of course, a problem, but not one that Bikinians feel is theirs, although this is the solution they favor. The other proposal is to treat the soil with potassium fertilizer on the theory that the plants will absorb it instead of radioactive materials. Tomaki Juda, son of a high chief, represents his people as notably unimpressed by this plan.

21. The official explanation for the natives' exposure was an unantici-

pated shift in the wind. Some Marshallese natives believe that they were not warned or removed for strategic reasons—so that the United States could use them as experimental subjects (Horwitz 1991).

22. Americans have not been the only ones to inscribe their order on the islands. The French exploded forty-one above-ground nuclear tests between 1966 and 1988 before extending the underwater order through detonations in several atolls in French Polynesia (Seager 1993: 61). Three, or in some reports five, more blasts tore apart more of Mururua Atoll in early 1996. The British tested nuclear weapons at Christmas Island but saved most of their testing for aboriginal lands in Australia (Seager 1993: 61). In all, the United States carried out 106 of the 312 atomic tests carried out by the three countries in the Pacific between 1946 and 1996 (Davis 1996: 8–9).

23. As we have done with the spelling of Hawaiian names, we have tried to honor the spellings of indigenous peoples. The English spelling follows in parentheses.

24. See Hinsdale (1988) for a discussion of the accession of land to the national government by states and through treaties with and purchase from Indian tribes. Johnson (1976) discusses the history of the rectangular survey order inscribed on much of the land of the United States. Hibbard (1965) discusses the formation of the public domain.

25. Three evolutionary stages of the development provided first for "total governmental authority in an appointed Governor," with an elected legislature and courts to be added as a second stage, before the final stride to statehood and popular government (Leibowitz 1989: 6).

26. The insular cases are *Downs v. Bidwell*, 182 U.S. 244 (1901); *Armstrong v. U.S.*, 182 U.S. 243 (1901); *Dooley v. U.S.*, 182 U.S. 222 (1901); *DeLima v. Bidwell*, 182 U.S. 1 (1901).

27. In 1947 the United Nations Trusteeship Council, in their *Trusteeship Agreement for the United States Trust Territory of the Pacific Islands*, indicated that the territories were specifically not an "integral part" of the United States. The trust territories were to be administered under the general clause directing trustee nations to foster economic, social, and political development, including whatever form of political development "might be appropriate to the particular circumstances of the trust territory and its people and the freely expressed wishes of the people concerned" (quoted in Wilson 1995: 25). In the Solomon Report, the United States worked out its strategic plan for circumventing inconvenient UN requirements about Micronesian people's "freely expressed wishes" and securing control over the valuable unclaimed real estate the United States saw in the Pacific.

28. American security logic has gone from denying the Micronesian islands to "the enemy," to retaining them as "fallback bases," and, at present, to maintaining them as forward positions (McHenry 1975: 69).

29. Although no terminal date had been set for ending the trustee status of Micronesia, worldwide independence winds were blowing through other former colonies and reaching Micronesia and tweaking American political leaders as well. This and the increasing recognition and criticism of America's miserable record in the social, economic, and political development of Micronesia to the clear benefit of U.S. military goals goaded the United States to alter the trustee relationship (McHenry 1975: 7).

30. Our account owes much to the detailed analysis in Wilson 1995. See also Roff 1987; Clark and Roff 1987. For a thorough legal and historical discussion of whether the UN should terminate the U.S. Trust Territory of the Pacific, see Clark 1980.

3. CONSTRUCTING AND CONTESTING THE FRAME AT FORT DERUSSY

1. We have both drawn on and taken some liberties with Roxanne Lynn Doty's (1996a) fascinating analysis and her inspired application of Homi K. Bhabha's conceptual framework to state policy on illegal immigrants.

2. Reported by Joelle Mulford in an interview with Capt. Timothy M. Johnson, Recruiting Operations Officer, Army ROTC, University of Hawai'i, January 31, 1991. See also Ali, Ferguson, and Turnbull 1991.

3. Such efforts to alter the terms within which security is constructed are being made by Cohn (1987), Enloe (1993), Tickner (1992), Peterson (1992), Sylvester (1994), Elshtain (1987), Campbell (1992), Shapiro and Alker (1996), and others engaged in the project of rethinking security.

4. It is worthwhile to recall his words: "Every gun that is fired, every warship launched, every rocket fired signifies, in the final sense, a theft from those who hunger and are not fed, those who are cold and not clothed. The world in arms is not spending money alone. It is spending the sweat of its laborers, the genius of its scientists, the hopes of its children (quoted in Moore 1995: 37).

5. Information about the United States' expanding role as international arms merchant is not widely disseminated. The media-watch program Project Censored claims that "America's role as arms supermarket to the world was the top underreported news story of 1997" (Zoll 1998: 6). The student-faculty program at Sonoma State University sees censorship operating in the United States, not by virtue of government intervention but as a result of corporate control of media: "The country's 11 major media corporations—which control the majority of the news Americans see, hear, and read—had 155 directors on their boards in 1996. Between them, these 155 people also hold 144 directorships on the boards of Fortune 1000 corporations in the United States" (Zoll 1998: 6).

6. It is difficult to keep track of the players in the mad whirl of mergers. Lockheed Martin has bought Northrop Grumman with a projected loss of ten thousand jobs while top company executives take millions of dollars in bonuses from public coffers (Vartabedian 1997: B-8). The fifteen members of the European Union initially opposed Boeing's merger with McDonnell Douglas on antitrust grounds, then concurred after receiving concessions from Boeing (Andrews 1997).

7. Charles Sennott's informative series of articles in the *Boston Globe* both reveal and conceal some of the performative energies at work in supporting the national security discourse's dangerous pedagogy. On the one hand, the articles relentlessly sniff out the alibis for the arms trade that are manifested in elements of political economy. On the other hand, the articles participate vigorously and without comment in the patriarchal dimensions of arms accumulation, manufacture, and sale. Sennott's prose reproduces fully and faithfully the phallic insecurities of countries with small missiles (e.g., "China: a nation with something to prove" [B-9]), the confident masculine excitements of the well-endowed states (e.g., "America's the muscle in an all-star cast" [B-3]). He is appropriately and gratifyingly suspicious of the self-interested justifications given by corporations and governments for obscene expenditures on weapons but entirely unaware of the masculine investments that also underwrite the arms trade. Sennott has a keen eye for the discourses of economic privilege and state power, but he has little visual acuity when it comes to phallocentric entitlements. Our thanks to the students in our Fall 1996 Feminist Theory class for their help in deciphering the mixture of speech and silence in Sennott's work.

8. See Soguk (forthcoming) on immigration policies as struggles over boundary negotiation; and Connolly (1991: 207–8) on the production of terrorism.

9. Allegations about illegal CIA involvement in arming the Nicaraguan Contras with funds gained by running drugs have been asserted and denied for a long time. The Christic Institute brought this story to the public in their November 1987 special report "The Contra-Drug Connection." The *San Jose Mercury Press* brought the story back into the public eye in Gary Webb's controversial 1996 series of articles "The Dark Alliance." Webb calls the CIA drug deals "a bizarre drug alliance," but what if this or other similarly illegal activity is standard operating procedure? (See also *Honolulu Advertiser* 1996d: A-8.) The combination of secrecy and paranoia that seems to characterize the CIA, and its privileged immunity from most congressional scrutiny, invites as well as hides such activities. For further discussion of the CIA's dubious contributions to national security see Draper (1997).

10. An earlier decision in *Kalipi v. Hawaiian Trust Company* 656 P.2d

745 (1982), along with the PASH decision, shows the state courts as recognizing certain long-standing Native Hawaiian claims. See MacKenzie 1991.

11. Governor Cayetano made this remark at a reception for the Okinawan Women's Peace Caravan, a group of women traveling around the United States to raise public awareness of the impact of the immense U.S. military presence in Okinawa. The reception was held at Washington Place (the governor's mansion), Honolulu, February 14, 1996.

12. One of the sexual laborers interviewed by Keahiolalo commented that the military men were more likely to frequent the red-light districts outside of Waikīkī, perhaps because of higher prices in Oʻahu's tourist mecca (Keahiolalo 1995). Michael Shapiro's interviews with women working the streets of Waikīkī also found that tourists, usually Japanese, are more frequent customers there than are military men (Shapiro 1997).

13. *Baehr v. Lewin* 852 P.2d at 54 (1993). In May 1993 and again in December 1996, the Hawaiʻi Supreme Court ruled in favor of three same-sex couples who had brought suit against the state in protest of its refusal to grant them marriage licenses. For an informative discussion of the implications of this case for gay and lesbian politics, see O'Donnell 1996b.

The Hawaiʻi Supreme Court has proven more liberal than the legislature, which fled from full endorsement of gay marriage into a more ambivalent Reciprocal Beneficiaries Act. Passed on July 8, 1997, this compromise act provides some, but not all, of the rights and obligations of marriage to adult couples who could not otherwise legally marry. On the same day, the legislature passed House Bill 117, which proposes a constitutional amendment to be placed on the 1998 ballot allowing the public to decide if the Hawaiʻi State Constitution should be amended to give power to the legislature to restrict marriage to heterosexual couples. Despite this double talk, Hawaiʻi is nonetheless the first state to have some form of statewide domestic partnership (Jenkins 1997).

14. Whether anyone except Native Hawaiians can be called local is also a contested notion. There was brisk exchange over this notion at the first International Multi-Ethnic Literature conference held in Honolulu in April 1997 and at the Hawaiʻi Literature Conference, March 12, 1994.

15. A small and unscientific sample of Hawaiʻi residents interviewed for a news story generally answered the question "Should the military's presence in Hawaii be increased or decreased?" by recognizing its economic importance but thinking the current size was about right (Matsumoto 1996: A-1).

16. There are important distinctions between the ways in which the regular army and the Hawaii National Guard relate to Hawaiʻi. According to General Eugene Imai of the Hawaii Army National Guard, the active-duty personnel come into an area and take it over, while the guard connects to

and works with the local community. General Imai noted that he can tell, just by watching how they do things, which personnel are guardsmen and which are regular army (interview with Imai 1995).

17. According to Sam Mitchell, president of the Machinists and Aerospace Workers, Local 1998, there are approximately two hundred federal employees from the mainland working at Pearl Harbor Shipyard. While this is a small number, it is nonetheless "a major issue with the guys here" (interview with Mitchell 1997).

18. The Damon Tract riot was not the first incident of this kind. In January 1826, the first U.S. warship to visit the islands sailed into Honolulu commanded by Lieut. John Percival, who was under orders to investigate debts owed by the chiefs to U.S. traders. He had little effect on that issue, but his advocacy of prostitution and his crew's rowdiness had considerable effect on the local market in sexual labor. Sailors demanded women and began to break up the home of a Christianized chief where church services were about to begin. Percival and his officers got control of their men; at the same time the commander put further pressure on the government to lift its *kapu* on women; eventually, Governor Boki yielded (Daws 1968: 78–79).

In 1916, a riot by troops in Iwilei "resulted in temporary martial law and temporary closing of the district" (Tavares 1945: 2). A similar upheaval occurred again in Honolulu some five years later (Tavares 1945: 2). Like his predecessor Lieut. Percival, the commander of the sailors involved in the Damon Tract riots faced court-martial and was acquitted (*Honolulu Star-Bulletin* 1946: 2).

19. For example, a recent newspaper article worried about local attacks on military personnel: "Sailor Attacked at Waimea Bay," *Honolulu Advertiser* 1996c: A-3.

20. While there are general patterns of appropriation and erasure in the military's institutional presence in Hawai'i, the attitudes of individual servicemembers are no doubt more varied and in any case difficult to document. Evidence is mostly anecdotal. In 1977, during the height of publicity on the Kaho'olawe protests, two marine pilots from Kāne'ohe Marine Corps Air Station responded to the Hawaiian activists by printing and distributing shirts saying "Bomb Kaho'olawe 'Ohana—Firepower for Freedom" (*Honolulu Advertiser* 1977: A-3). In the Hawaiian language, 'ohana means a family or kin or group; the Protect Kaho'olawe 'Ohana was formed in the 1970s to protest the navy's bombing of the island and to begin to pressure the federal government to return the island to the state and, ultimately, to the control of Native Hawaiians. The pilots later claimed that the shirts were in jest.

On the other hand, Mehmed Ali, then a Marine corporal, related this story: "Manny Kuloloio told me of the time the Protect Kaho'olawe 'Ohana printed some green camouflage "Stop the Bombing of Kaho'olawe" T-shirts.

When they went to Kahoʻolawe, they gave the T-shirts to the military personnel working there, who all wore the shirts. When the commander flew over
in his helicopter, he saw all the troops wearing the shirts and ordered them
banned" (interview with Ali 1991). Perhaps this was also in jest. We do not
know. Our point is that our analysis of the semiotic and institutional presence of the military in Hawaiʻi cannot adequately represent all individual
servicemembers, who no doubt exceed and confound as well as conform to
official expectations. There are examples like that of U.S. Navy Commander
Joel Keefer, who noted to us with some regret that military personnel tend to
trivialize Native Hawaiian culture, and that the overthrow was "not a thing
we should be terribly proud of" (interview with Keefer 1991). A more radical voice of protest came from Jeffrey Paterson, a young marine stationed in
Hawaiʻi, who was the first public military resister to refuse deployment
in the Gulf War. His case was covered extensively in the media until he was
finally court-martialed and then given an administrative discharge (Ensign
1992: 286). Locally, Paterson's case was a rallying point for opposition to
the Gulf War. On the other hand, during a widely publicized conflict at the
University of Hawaiʻi at Mānoa in 1990–91 over antihaole remarks made
by a Hawaiian studies professor and sovereignty leader, Haunani-Kay Trask,
a different attitude was expressed in the considerable number of letters to
the editor in the local papers that were from military personnel. In one such
letter, Sergeant Major Robert R. Rogers (U.S. Army) stated: "The people
who profess to be the rightful owners of the land have much less respect
for it, and themselves, than any other group I have ever seen. I agree with
Professor Trask—the whites should leave Hawaii. They should take the
military, tourism, and statehood and leave this rock to the ocean; and the
people will turn it into the largest garbage dump in the world" (*Honolulu
Advertiser* 1990b: A-15). It is difficult to say with certainty that Major
Rogers's remarks are more representative of servicemembers' views than are
those of ex-Corporal Paterson or Commander Keefer, but it does seem likely.

21. "Haolefied" is based on the Hawaiian word *haole,* which means outsider. It has taken on a specific reference to Caucasians, and here it means
that the original Hawaiian names were changed to English or transliterated
into English with little or no reference to their former names, meanings, or
spellings. The word *haole* is a good example: it has become a common, if
contested, word in English. For an excellent discussion of the subject see
Rohrer 1997.

22. The only fishpond still in existence is in the waters of West Loch, inaccessible to any civilians because of its location near the Naval Magazine.

23. For a brief description of the main sovereignty groups, see Omandam
1995b: A-1. See also Castanha 1996.

24. The history of Native Hawaiian resistance to the military in particular

and to colonization in general is extraordinarily complex. Much of the opposition has been covered only in the native language press and still receives little attention in English sources. While this history resists adequate summary, a few prominent instances can be named.

With regard to statehood, Alice Kamokila Campbell, a part-Hawaiian territorial senator in 1942, was a prominent opponent of statehood. While it is unclear how representative her views were among her people, Roger Bell notes that "Hawaiians and part-Hawaiians were more likely to oppose statehood than were members of other ethnic groups" (Bell 1984: 116). Our thanks to Norm Meller for sharing his knowledge of the politics of statehood.

With regard to the dominance of foreigners in the monarchy, in 1889 a group of Native Hawaiians, led by the charismatic figure Robert Wilcox, revolted against the Bayonet Constitution, as it came to be called, that had been forced on King Kalākaua by the haole elite. This constitution had given U.S. citizens the right to vote in Hawaiian elections, while many Hawaiians were excluded because of property restrictions on suffrage. The constitution had also taken from the king the right to appoint and dismiss his cabinet members. Wilcox's rebellion was put down by U.S. Marines from the USS *Adams*, who "distributed ten thousand rounds of ammunition to government forces" to guarantee against future protests (Kent 1983: 59; see also Kuykendall 1967: 424–30).

With regard to resistance to the efforts to secure a reciprocity treaty, and to annexation, Kuykendall notes that despite foreigners' enthusiasm, "it is doubtful . . . that very many Hawaiians were converted to the cause of annexation" (1966: 221, 226–27). Daws gives a brief account of royalist efforts to restore the queen (1968: 281–85). The forthcoming dissertation by Noenoe Silva promises a richly documented history of organized, widespread Hawaiian opposition to Annexation. Based on Hawaiian-language newspapers of the time, "*Ke Kūʻē Kūpaʻa loa nei Mākou*: A Study of Hawaiian Resistance to Colonization" will substantially revise English-language analyses of Hawaiian history. See also Silva 1997.

25. For an interesting discussion by a Native Hawaiian woman on the dilemmas of pursuing sovereignty without essentializing identity, see O'Donnell 1996a.

26. In an article in the local Kauaʻi paper Anne E. O'Malley interviewed Sharon Ann Mapuana Pomroy about her opposition to the U.S. Army plan to launch missiles from Barking Sands in Kauaʻi as part of its STARS development program (1990: C-1, C-7). Citing environmental dangers and the destruction of Hawaiian archaeological sites, Pomroy demands, "If they think this thing is safe, and it's not going to cause so much harm, then why the hell don't they launch it in New York City? Why don't they clear a spot on Fifth Avenue and blast one off?" (C-1).

27. We use the vague phrase *large parcel* because it is difficult to get an accurate measure of this land. A 1990 article in the *Honolulu Advertiser* cites Charles Maxwell, vice chair of the Hawaii Advisory Committee to the U.S. Civil Rights Commission, as asking for the return of thirteen thousand acres at Lualualei (in Borg 1990). In 1995 the Department of Defense issued its "Draft Hawaii Military Land Use Master Plan," which identifies about eleven thousand acres of military land for potential "sale, exchange or release," including seven thousand acres of the munitions facility at Lualualei. Much was made in the local papers of the impending return of Lualualei, with Senator Inouye quickly endorsing the plan. The draft report estimates that the process will take fifteen to twenty years, so the celebrating may have been a bit premature. See Rohrer (1995: 12–14). While Judy Rohrer's persistent efforts to pin down the land figures were unsuccessful, her thoughtful analysis of military occupation of land has been crucial for our own understanding.

28. Zohl de Ishtar recounts the experience of a Hawaiian woman whose family was warned to flee from their garden at Waikīkī to the hills to escape a tidal wave. The warning was a ruse, for when they returned, dredged material from an engineering operation to "reclaim" the Waikīkī wetlands covered their land (1994: 102). Barry Nakamura's history of this "reclamation" operation is a careful account of the processes by which wet agriculture and aquaculture in Honolulu were condemned as unhealthy. The solution proposed and executed was to drain the fertile lands through dredging and filling. Authorized by the territorial government as a health measure, the dredging process itself was especially lucrative for Walter F. Dillingham, who had in the several preceding decades bought several large parcels of the wetlands, including the 84.41-acre site later developed as the Ala Moana Shopping Center, the Ala Moana building, and the Ala Moana Hotel (Nakamura 1979: 70).

29. Both the Environmental Protection Agency and the State Department of Health investigated the high incidence of childhood leukemia among those living near the transmitter, but in the end the reports were inconclusive. While the newspaper story on the department study proclaimed that "no leukemia link [was] found to Lualualei," the woman who ran the investigation said that it was "inconclusive" because the numbers were statistically insignificant, while also admitting that "the sick children lived somewhat closer to the towers than the healthy children" (Altonn 1992: A-1, A-8).

30. On May 14, 1996, a twenty-three-mile underground pipeline at the Waiau power plant leaked, undetected, for several hours, spilling thirty-nine thousand gallons of fuel oil into Pearl Harbor. As of early 1997, the financial cost of this spill was $1.2 million for the federal government and $10 million for Chevron (Billig 1997: 16–17).

A lengthy study by the University of Hawai'i Sea Grant program of actual and potential oil spills in Hawai'i documents the billions of dollars in financial and environmental costs that would follow from a major spill. With regard to Pearl Harbor, the report voices concern about the undocumented environmental damage from the aged infrastructure and "continuous low-level oil spills" (Pfund 1992: 115). The report calls for regular maintenance and replacement of equipment and facilities, and "rigorous risk management and loss prevention programs," which could reduce or eliminate the 40 percent of spills caused by mechanical/structural failure. The report concludes that "the mystique of a pristine environment which lures visitors to pay higher transportation costs to stay in upscale beachfront resorts would be shattered, if the news media were to transmit worldwide views of Hawai'i's coastal vistas coated with black oil" (Pfund 1992: viii).

31. Individual cases of violence toward women and children in military families are frequently reported in the local newspapers. State officials maintain that 5.5 percent of reported child abuse and neglect cases involve military personnel (Hawaii Department of Human Services 1994: 23). Given the negative effect that official reports of family violence may have on a soldier's career, it is likely that many incidents go unreported. Vicki Nielsen indicates that children in military families are nearly twice as likely to have alcoholic fathers as are civilian children, a factor that probably increases violence in military homes (Nielsen 1996: 1). Women's groups such as the Joint Military Family Abuse Shelter and the Navy Assault Victim Intervention Program have organized on O'ahu to address domestic violence in military families.

32. Title to Kaho'olawe was transferred to the State of Hawai'i, pending its transfer to a "sovereignty authority." Several Hawaiian groups are competing for the authority.

4. REMEMBERING AND FORGETTING AT PUNCHBOWL NATIONAL CEMETERY

1. The cemetery is administered by the National Cemetery System of the Veterans Administration, while the memorial was erected by the American Battle Monuments Commission. The commission was officially chartered in 1923 to "'control,' 'censor' and monitor battle monuments and cemeteries as 'proper' representations of the United States abroad" (Robin 1995: 57).

2. For a thorough discussion of Punchbowl's history, see Nogelmeier 1985, esp. 22–23. See also Kamakau (1991: 25–26).

3. Traditional Hawai'i had an oral culture; in such cultures, meanings of words are stored and retrieved differently from the ways they are in a written culture. For an expanded treatment, see Buck 1993. Moreover,

meanings are heavily dependent on the context and the relationships of the people speaking and what they are talking about. Birds are not always birds and waterfalls are not always waterfalls. It is commonly said that if the song is about a canoe and ocean spray and Hawaiians are laughing, the song is not a sea chanty.

4. For example, lower-ranked persons were forbidden to allow their shadows to fall on the person or shadow of the queen. We thank Nahua Patrinos for this understanding and for the example. See also Kamakau (1964: 9–10). The religious and *kapu* systems were gendered in that women were excluded from certain activities, separating "the sacred male element from the dangerous female, thus creating order in the world" (Kame'eleihiwa 1992: 25). Women were forbidden certain foods due to their male symbolism (pig, coconut, bananas, certain red fish); they ate apart from men, as Lilikalā Kame'eleihiwa explains: "For women to eat these foods would not only allow their *mana* [spiritual power] to defile the sacrifice to the male *Akua* [god] Lono (*Akua* of fertility and agriculture), Ku (*Akua* of war and wood carving), Kanaloa (*Akua* of the ocean and ocean travel), and Ku'ula (*Akua* of deep-sea fishing). Given that the word *'ai* means "to eat, to devour" and also "to rule and to control," if women ate the *kinolau* of these *Akua,* they would gain the *mana* to rule the domains represented by these *Akua*; women could then rule male sexual prowess, including war, agriculture, ocean travel, and deep-sea fishing. What would be left for men to do?" (Kame'eleihiwa 1992: 34).

Genealogy, on which rank depended, was traced bilaterally. Documents demonstrate that chiefly women were powerful actors. For an excellent discussion, see Linnekin 1990. Jocelyn Linnekin also mentions a study dealing with unsuccessful missionary attempts to teach "submissiveness" to Hawaiian women. Kame'eleihiwa emphasizes the strength and power of women in traditional Hawaiian literature and rejects the common anthropological interpretation of *kapu* as male dominance: "As a Hawaiian woman, I can frankly say that I would not mind if I never ate with a man. I can think of many more interesting things to do with men than to eat with them. And, if it meant that men would do all the cooking and that only men would be offered in sacrifice, I would, like my ancestral grandmother Papa, agree to this law. I do not find the *'Aikapu* demeaning. Instead, it seems likely to me that under this religion men in traditional Hawai'i worked harder than men do in modern Western society. Perhaps Papa felt as I do and decided that eating without men and forgoing certain foods [were] well worth the exchange of men doing their share of the work!" (Kame'eleihiwa 1992: 36).

5. The practice of burning the bodies of *kapu* violators is said to have come to O'ahu from Kaua'i around the fifteenth century (Nogelmeier 1985: 35).

6. *Kalo* is the Hawaiian name for the tuberous plant that was a staple of the traditional Hawaiian diet and is still an important foodstuff in many parts of the Pacific. The Tahitian name is *taro*.

7. The Hawaiian word *kapu* and the Tongan *tabu* are both rendered in English as *taboo* (O'Donnell 1996a: 22).

8. American space does not have quite the sense of possession that American soil has. American soil is a particular space that has been watered with the blood of martial patriots and the sweat of agrarian patriots. The transgression of American airspace conjures up thoughts of different kinds of Patriots.

9. See Carlson (1990) for a longer history of its creation.

10. Military policy after World War II was to repatriate the remains of troops buried in "temporary" cemeteries in the Pacific. The next of kin were given the choice between having the bodies shipped to hometowns or national cemeteries for burial (from information provided by the director of the National Cemetery of the Pacific in an interview, July 25, 1994). Of the more than 13,000 buried, 11,597 were identified and 2,079 unidentified. They came from Guadalcanal, China, Burma, Saipan, Guam, Iwo Jima, and prisoner of war camps in Japan. There is another National Memorial Cemetery in Manila where other Pacific war dead are buried (from an information sheet prepared by the 100th Infantry Battalion Club, available at the cemetery office).

11. Because it is not the direction of this chapter, clearly this account of burial sites and practices in America is thin. We are aware of the many exceptions to the churchyard mode: mass graves of victims of disease; the wars of extermination of Native American Indians; the unmarked graves of slaves; the individual graves of those who perished in the westward movement across North America; the graves of Civil War dead; the quite different Amerindian practices dealing with the dead. Jackson and Vergara (1989) graphically document the varieties of cemeteries and their changes through time.

12. The chairman of the Pacific War Memorial Commission, in a report to the 1962 Hawaii State Legislature, noted that the anticipated dedication of the Gardens of the Missing monument (largely federally financed) would stimulate tourism: "With more than 25,000 names of Pacific Area combat missing-in-action engraved in marble—representing all sections of the Nation—it is anticipated that the influx of visitors to the dedication ceremonies may conceivably mark an event of major proportions in Hawaii's tourist history. . . . In the years to come, it is anticipated that the creation of the "Gardens of the Missing" Monument will significantly expand the current visitor level of approximately 300,000 to Puowaina" (1962: 51). (The area is now commonly referred to as Courts of the Missing.)

13. Not all of the more than thirteen thousand came from the Pacific

Islands. Some were from mausoleums and cemeteries in Hawai'i and on the mainland. See Carlson 1990: 35, 41.

14. Estimate by the Cemetery authorities, August 1994.

15. At an earlier time, fresh flowers could be placed on graves only through florists that were members of the Florists Transworld Delivery (FTD) system (Singletary 1977: 5).

16. This story was shared by a student in our Fall 1995 Learning Community, Women and Men, Peace and War class, Department of Political Science, University of Hawai'i, Mānoa.

17. All quotations that we cite in this section are from the mosaic texts and illustrations at the memorial.

18. For a useful overview of Hegel's political philosophy, see Sibley (1970: 434–61). For useful overviews of Hegel's views on war, see Avineri (1972: 194–207); and Dallmayr (1993: 155–59).

19. These comments come from a conversation we had with three veterans and their wives at Punchbowl Cemetery, October 21, 1993.

20. In *Leviathan,* Thomas Hobbes noted the connections between war and the pervasive inclinations of a society: "For WARRE, consisteth not in Battell onely, or the act of fighting; but in a tract of time, wherein the Will to contend by Battell is sufficiently known: and therefore the notion of *time,* is to be considered in the nature of Warre; as Foule weather, lyeth not in a showre or two of rain; but in an inclination thereto of many dayes together: So the nature of War, consisteth not in actuall fighting; but in the known disposition thereto, during all the time there is no assurance to the contrary. All other time is peace" (1968: 185–86).

21. One wonders what visitors from other nations, specifically from Japan, make of this grammatical draft, from which they apparently hold exemptions.

22. Thanks to Katharina Heyer for helping us learn about Japanese women's lives during World War II.

23. "I even recollect having read that a clergyman used to ring a bell at midnight to remind them to perform their matrimonial duties, for it would otherwise never have occurred to them to do so" (Hegel 1975: 154). Here, Hegel is talking about Native Americans; he does not see them as "lascivious," compared to their African counterparts.

24. For example, recent photos in Honolulu newspapers showed local women performing the hula for sailors returning to port after a long deployment at sea.

5. SEEING AS BELIEVING AT THE *ARIZONA* MEMORIAL

1. John Berger (1979) shows how all seeing is socially framed. Martin Jay (1994) writes a history of visuality and the discourse of "ocularcentrism."

Kenneth Little's work (in Howes 1991) is a critique of the safari as a cultural production and its relation to the "tourist perspective." See also Timothy Mitchell (1991).

2. More than six million tourists visited Hawai'i in 1993; the state's population is about one million (Hawaii Department of Business, Economic Development and Tourism 1994: 11, 169).

3. The "world as picture" is from the explanation by Heidegger (1977) of the formation of the modern objectified view of the world.

4. A small sample of other accounts of the war includes John Dower (1986), whose work deals with the intense racism on both sides, making the Pacific War a vicious and savage conflict that could only end through unconditional surrender. Casualties were high on both sides, and few prisoners were taken. Dower also discusses how this commitment to an all-out victory, combined with the new technologies of destruction, changed war's nature and made possible the killing of masses of people, particularly civilians, at great distances. Gerald Wheeler (1963) offers a history of Japanese–United States relations in the two decades prior to the attack on Pearl Harbor. He argues both that war in the "Far East" was regarded as likely by U.S. naval officers during this time and that Japan was regarded as the likely enemy in the Pacific by the War Plans Division and the Naval War College.

In his work, Akira Iriye (1972) traces the estrangement of Japan and the United States during parallel periods of imperialism by both nations. Other investigations and hearings focused on the attack and military vulnerability. Between 1941 and 1946, eight separate investigations into the attack were held by various U.S. authorities including the president, the Secretary of the Navy, the Secretary of War, and, after the war, a Joint Congressional Committee. There were thirty-nine volumes of testimony and evidence; the majority and minority reports of the Joint Congressional Committee, which logically make up the "fortieth volume," were cataloged separately from the hearings (see United States Congress 1976). None of the forty volumes was either cross-referenced or indexed as a unit until Smith (1990).

Some sense of the politics of military promotion and command is contained in Admiral James O. Richardson's memoirs, interestingly titled *On the Treadmill to Pearl Harbor* (1973). President Franklin D. Roosevelt relieved Richardson of his post as commander-in-chief, U.S. Fleet (CINCUS), only thirteen months after appointing him; the normal tour would have been eighteen to twenty-four months. Richardson attributes his removal from command to his having differed with FDR about the imminence of war with Japan and the need for more arms and men at Pearl Harbor. Richardson faults Chief of Naval Operations Admiral Harold Stark for not having vigorously enough prosecuted plans for defending military forces in Hawai'i just before the attack. He also defends Admiral Husband Edward Kimmel, on whom the blame was heaped, and condemns Stark for not supporting

Kimmel. Of Stark's general abilities, he says: "I do believe that the United States Fleet would not have been in Pearl Harbor on December 7, 1941, had I been Chief of Naval Operations at that time" (10).

William Iles (1978) argues that from the outset authorities sought to locate blame, not responsibility. President Roosevelt continued in this direction when he instituted the first inquiry, which resulted in charges of poor judgment and dereliction of duty against General Walter Short and Admiral Kimmel. They were retired from service. Iles also shows that subsequent investigations shied away from institutional explanations that would have framed responsibility as the issue.

5. We have borrowed this felicitous phrase from Mary Louise Pratt (1992: 201) and added company to its original first-person singular.

6. See Coffman (1973: epilogue II); see also Lind (1992: 1). When H-3 is completed, the state will have a total of 51.5 miles of interstate highways. Information from an interview with a State of Hawai'i Department of Transportation spokesman, November 29, 1996.

7. The location of the *Missouri* was the subject of extensive legislative dispute. In July 1997, the Congress selected Hawai'i as the winner, enabling the ship symbolizing the end of World War II to be berthed next to the ship symbolizing its beginning. Amid the navy's enthusiastic plans to "expand and coordinate military tourism in a major way at Pearl Harbor," the National Park Service is concerned about maintaining appropriate solemnity and avoiding a "theme park" atmosphere (Griffin 1997: B-1).

8. Rarely visible, some of these politics is suggested by Peter Boyer's discussion of the career and suicide of Admiral Jeremy Boorda (1996: 68–86). He came up to the command level through the ranks, not from Annapolis. His appointment as Chief of Naval Operations rankled many senior officers, who preferred a more traditional candidate. Another inside view emerged in the air force announcement on May 30, 1996, that three senior officers had been relieved of their commands following the investigation of the crash in April 1996, in which Secretary of Commerce Ronald Brown and all thirty-four others aboard an air force plane were killed (*Los Angeles Times* 1996a: A-l). The air force conceded that the crash could have been averted had the pilots and their senior officers followed safety rules for instrument landings at underequipped airports (*Los Angeles Times* 1996b: A-1).

9. Copies of a free brochure, "Official Map Guide to Oahu's Premier Attractions," published by the Oahu Attraction Associations, are available. The *Arizona* Memorial is listed as one of the twenty "premier attractions."

10. At one time, koa covered about two hundred thousand acres on the island of Hawai'i (Hartwell 1985: A-3). In 1985, there was only one healthy, virgin koa forest left in Hawai'i. Present-day people of Hawaiian ancestry

explain that their ancestors planted a single tree at a time amid a diverse ecosystem (*Honolulu Star-Bulletin* 1985: 12).

11. The National Park Service and the U.S. Navy manage the memorial, but because it is a major national memory site, many other voices affect the frames. These include the *Arizona* Memorial Museum Association, veterans' groups, Pearl Harbor veterans, and volunteers at the memorial. Exhibitions at the museum are dynamic; their changes over time reflect not only these various voices but findings of staff research, fluctuating funding, and such cultural transformations as increasing numbers of foreign visitors and the younger ages of American visitors.

12. C. D. Bryant (1976: 129) gives an account of the experiences of the Iowa farm parents of Sgt. Michael Mullen, who were determined to break up the unitary story surrounding their son's death in Vietnam. Their experience of being long stonewalled by the Department of Defense jarred loose the legitimacy of the state's hold on them. The Mullens spent the gratuity payment from the army (meant to pay for his funeral) on a half-page ad in the Sunday edition of the main Iowa newspaper, whose half-inch-high headline read, "A silent message to fathers and mothers of Iowa." Beneath ran fourteen rows of forty-nine crosses each, and a fifteenth row with twenty-seven, leaving blank space for more. Each of the 713 crosses represented the death of an Iowa soldier in the Vietnam War.

13. Four hundred and sixty-nine Medals of Honor were awarded for acts of heroism during World War II; none was awarded during that time to African American men, although they had won them in all the wars prior to the Second World War and again in the two succeeding ones. In May 1996, President Clinton awarded seven Medals of Honor, six posthumously, to African American soldiers. Of Hawai'i's 442nd Regimental Combat Team and the 100th Battalion, which saw some of the fiercest fighting in Europe and were among the most decorated units, only Sadao Munemori of Los Angeles was awarded the Medal of Honor. Fifty-two members of the 100th Battalion and the 442nd Regimental Combat Team, including Senator Inouye, received the Distinguished Service Cross. Barney Hajiro of Pearl City, Hawai'i, was nominated for the Medal of Honor. His nomination and combat wounds kept him from further combat; when again fit for duty, he was assigned to one of the army's segregated (all-black) units. Hajiro, now eighty, does not know why the award was never made, but he suggests, "Maybe it was because I was Oriental" (Kakesako 1996b: A-1, A-6).

14. In common parlance, a ship is perpetually female, steered on a course set by a commander and worked by a crew that mans her.

15. Grosz concludes that in order for women to occupy space in a different way, space needs to be reconceived in ways that do not reproduce the "logic of penetration, colonization, and domination" (123). This suggests

that the military's program of including women within their ranks by "adding women and stirring" will not interrupt the (male) metanarrative that governs modern space.

16. The current film replaced one produced by the navy in 1980. Several reasons drove the decision to make a new one. The old one framed "the lesson of Pearl Harbor in terms of military strength" (White 1998: 733). With the fiftieth anniversary of V-J Day approaching and the Cold War having ended, it was thought that a less militarily aggressive story would be more appropriate. The present film was finished in 1993. See White (1998) for an extended discussion of the politics of the re-creation of the Pearl Harbor history.

17. On December 7, 1996, the ashes of two survivors of the *Arizona* were interred in the gun turret. The ashes of ten other sailors had previously been placed there (Kakesako 1997: A-3).

18. The following is taken from the dedication by Admiral A. W. Radford, USN, March 1950: "Dedicated to the eternal memory of our gallant shipmates in the USS *Arizona* who gave their lives in action 7 December 1941. From today on the USS *Arizona* will again fly our country's flag just as proudly as she did on the morning of 7 December 1941. I am sure the *Arizona*'s crew will know and appreciate what we are doing." Admiral Radford is correct that our country's flag flies over the site again, but he failed to anticipate the brevity of each proud unfurling. Two sailors periodically run American flags up and down the flagpole in order to restock the gift shop's inventory of flags that "have flown over the *Arizona*." It is not clear whether the crew either know or appreciate what they are doing.

19. Not only were the fishponds and fish lost but so was a large freshwater lake into which "metal scrap, engine parts, empty ammunition casings, and airplane and ship parts" were dumped. Military equipment stored in marsh areas, awaiting return to the mainland, sank. "Petroleum pipelines were laid and abandoned almost at will. Unlined pits that drained straight into the harbor received everything from petroleum sludges to paint to battery acid. Storm drains provided a convenient means of disposing of everything from mercury to chromium" (Tummons 1991e: 7, 8). See also Tummons 1991a, 1991c, 1991d.

20. Barthes (1978: 124) explains that "myth is a type of speech defined by its intentions . . . more than its literal sense."

21. Some bits of the rusted wreckage of the USS *Utah*, which capsized after being torpedoed, are visible on the side of Ford Island opposite the *Arizona* Memorial. A small memorial to the *Utah* was built on land in 1971; the wreckage, otherwise unmarked, entombs fifty-eight members of its crew. The *Utah* was a training ship, not a combat vessel, at the time of the attack (Lenihan 1989). After the war, two other damaged and refitted battleships,

the *Nevada* and the *Pennsylvania*, became part of the "ghost fleet" (Delgado 1996), targets for the atomic bomb tests at Bikini. The *Nevada* did not sink, but, heavily contaminated, it was towed back to Pearl Harbor where efforts to clean it (and other vessels) proved unavailing. In July 1948, it was towed sixty-five miles south of Pearl Harbor and sunk in "water five miles deep" after four and a half days of bombardment by ships and torpedo bombers (Delgado 1996: 93).

22. The men are officially declared buried at sea (Lenihan 1989: 178).

23. See chapter 2 generally.

24. The words to the song "Let's Remember Pearl Harbor" were written by Don Reid, with music by Reid and Sammy Kaye (Republic Music Corporation 1941).

25. Dower (1986) shows textually as well as pictorially not only the American demonization of the Japanese but the reciprocal Japanese gesture. Such practices undergird the conventional self-understandings of nation-states.

6. THE PEDAGOGY OF CITIZENSHIP

1. See Carol Pateman (1988) for a feminist critique of social contract theory. For feminist analyses that both draw on and take issue with Pateman, see Kathleen Jones (1993: 42–44, 58–59, 62–71); and Nancy Hirschmann (1992: 85–89, 121–22).

2. For a discussion emphasizing competing bourgeois and Protestant traditions within America's liberal patrimony, see Roelofs (1992).

3. Our concern is not so much with the adequacy of Kann's reading of Locke as with the efficacy of that reading for interpreting citizenship in the national security state. For a discussion of competing interpretations of Locke, see Kann (1991: 221–23).

4. For a fuller discussion see Okihiro (1991).

5. Japanese American National Guardsmen were removed from the National Guard. After guardsmen persisted in requesting service, they were "turned over to the Army engineers as a labor force" (Fuchs 1961: 305; see also Allen 1950). Fuchs (302) also recounts how members of the local oligarchy resisted the idea of mass internment because they knew the Japanese constituted the main workforce. See also Rademaker (1951: 44).

6. This had also been true for the issei parents of Inouye's generation. Of them he said, "They had wanted so desperately to be accepted, to be good Americans. And now, in a few cataclysmic minutes, it was all undone, for in the marrow of my bones I knew that there was only deep trouble ahead" (Aloha Council: 1988a). The issei were the first generation of Japanese immigrants to Hawai'i; their second-generation children were the nisei.

Born in Hawai'i, the nisei were also American citizens. Those whose parents had registered their births in Japan held dual citizenship until they chose or were required to renounce one.

7. In 1946, Inouye and Bob Dole were patients at Percy Jones General Hospital in Battle Creek, Michigan, where both spent months recuperating from their combat wounds. Dole remembered Inouye as the best bridge player there ("Inouye and Dole Go Back a Long Way," Honolulu Star-Bulletin 1996c: A-3).

8. For a fuller discussion of the formation and first twenty years of the Democratic Party in Hawai'i, see Cooper and Daws (1985).

9. The East-West Center is an independent institution situated on the Mānoa campus of the University of Hawai'i. Its mandate, adopted in 1960, "is to promote better relations and understanding between the U.S. and nations of Asia and the Pacific through cooperative study, training, and research" (East-West Center Public Relations Office). The center has been the site of periodic struggles over the institution's proximity to the State Department and Pentagon interests.

10. The mainland-based Citizens Against Government Waste singled out Senator Inouye as the "pig book oinker of 1996," suggesting that Hawai'i receives a disproportionate amount of federal funds (Dingeman 1996: A-1). We have shifted the porcine metaphors away from the common accusation of "pork," as in pork-barrel legislation, to that of "bacon," as in bringing home the bacon, to underscore the Senator's paternal role in taking care of Hawai'i.

11. Hawai'i has no indigenous snakes. The brown tree snake on Guam has nearly wiped out native bird life there. It is a fast-moving reptile that can whip up and down trees and utility poles with ease and has been known to bite small children in their beds. The snakes occasionally get carried to Hawai'i in the cargo bays or wheel wells of military planes, much to the alarm of Hawai'i residents.

12. The seamlessness of the Inouye narrative in Hawai'i has been interrupted only twice by countericonic claims. The first centered on a widely publicized accusation of sexual harassment, the second on a quickly hushed story of battlefield improprieties. Shortly before Inouye's November 1992 reelection, Republican opponent Rick Reed published television and newspaper ads accusing the Senator of sexually harassing his hairdresser, Lenore Kwock. Senator Inouye denied the charges. Reed's allegations were made public without Kwock's permission and were pulled later at her request (Tanahara and Yoshishige 1992: A-1). The most remarkable aspect of this incident is the silence with which it was greeted by all but a handful of public officials. While Senator Inouye experienced the publicity, which spanned a period of several months, as "living hell" (quoted in Borreca 1992: A-3),

only one prominent woman Democratic leader and two Republicans made significant public statements. Democratic State Representative Annelle Amaral, a former policewoman and well-known advocate of women's rights, received about sixty phone calls on the issue, including nine from other women who reported similar incidents but refused to identify themselves publicly (Gross 1992: C-10; Yamaguchi 1992: A-3). Amaral stated that she believed Kwock and was subsequently defeated for legislative leadership positions she had been expected to win (Gross 1992: C-10). Republican State Representatives Cynthia Thielen and Jane Tatibouet also made public statements, but as members of the nearly invisible opposition party they had relatively little to lose.

Compared to the prolonged exposure and serious consequences accompanying similar allegations against mainland political figures such as Senators Bob Packwood and Brock Adams, the silence surrounding the allegations against Senator Inouye was deafening. To some extent this relative quiet can be explained in terms of loyalty to the Senator himself and to the victories against racial prejudice that his career and his commitments represent. "Ms. Amaral, a native Hawaiian, wept as she described 'choosing between the gender battle and the ethnic battle' because Mr. Inouye has long helped the native people" (Gross 1992: C-10). A more institutional explanation looks to the consequences of one-party dominance in the state: the perceived danger to careers that accompanies any independent speech, and the lack of political space for viable alternatives, assures that "dissident voices remain at the margins because they have nowhere else to go" (Milner 1992: A-19). A measure of this political closure is an internal memorandum circulated to employees within a major state social service agency, which stated in no uncertain terms that "your agency has elected *not* to get involved in this issue" and required of workers that "you do not participate in these activities or make any statements, even casual, to media members of our community regarding this matter" (Memorandum to Program Directors and Administrators, Child and Family Services, November 19, 1992). Only Ian Lind's independent *Hawai'i Monitor* used the situation to criticize "Democrats and Republicans alike . . . [for] forming a human shield around the Inouye pork barrel" and to call for serious public attention to the issues of sexual harassment and exploitation. (*Hawai'i Monitor* 1992: 4). There is some evidence that the voting public was more moved than their leadership, since "preelection polls found that 42 percent of likely voters believed Ms. Kwock, who passed a lie-detector test, and 20 percent Mr. Inouye" (Gross 1992: C-10). Yet the Senator won reelection to his sixth term by a wide margin. Indeed, as staff counsel to Governor Cayetano stated at the time, "Senator Inouye is an institution" (quoted in Gross 1992: C-10).

The iconic narrative of soldiering and citizenship that is the public Dan

Inouye was interrupted briefly on one other occasion. At the opening banquet of the fiftieth anniversary celebration of the formation of the 442nd Regimental Combat Unit, an old friend and fellow soldier recalled that forty-nine years earlier Inouye had removed a silver ring from the body of a dead French woman after a battle near the French town of Bruyéres. Many of the more than twenty-six hundred guests, including veterans and their wives, military officials, and a delegation of French nationals from Bruyéres, were shocked by the grizzly inclusion in what some described as a "rambling introduction" (Kakesako 1993: A-8). Some listeners remembered the speaker saying that "Inouye used a trench knife to cut a diamond ring from the finger of a dead French woman" (Kakesako 1993: A-8). The speaker, Teruo Ihara, denied saying that Inouye had cut off the finger to get the ring. The Senator explained the incident as part of wartime horror (he wore the ring "just as a reminder of how beauty can be destroyed by war") and wartime irony ("I take the ring off of the woman's finger and then a doctor has to snip off my finger [of his amputated right arm] to get the ring off") (in Kakesako 1993: A-8). Inouye, himself stunned at the unexpected public remark, referred indirectly to the advancing age of the speaker: "It reminded me that all of us were getting older" (in Kakesako 1993: A-8). Ron Oba, president of the 442nd Veterans Club, called the story "in bad taste" and its public telling misplaced: "War stories like that should be kept to the confines of the hospitality rooms." Unlike the sexual harassment episode, the battlefield incident quickly disappeared from the local news after the initial story in the *Honolulu Advertiser* and a passing mention in the *Honolulu Star-Bulletin* the same day (Keir 1993: A-8).

Our point in remarking on these two incidents is threefold: first, to recognize that the mythic presence of Senator Inouye within Hawai'i's public spaces has been countered on rare occasions; second, to highlight the enduring mythic as well as institutional power that effectively limits these interruptions; and third, to note the mandated silences that accompany and enable the narrative placements of war and sex in patriarchal public life. In the eyes of most other public officials and the mainstream media, Senator Inouye had only to call Lenore Kwock a liar to make it so. In almost everyone's eyes, possible battlefield improprieties (but not, of course, battlefield heroics) were irrelevant to the intertwining of military experience and civic power. Perhaps as a society we would begin to understand war and power differently if stories of harassment and abuse were put at their center rather than "kept to the confines of the hospitality rooms" and the hush of private conversations.

13. Teachers in the Japanese language schools as well as Shinto and Buddhist priests were among those interned. The "nursing fathers" of the Japanese American community were thus removed from their positions as local

leaders and advisers, and "leadership in the Japanese community passed immediately to the group holding American citizenship, only a few of whom were more than 40 years of age" (Allen 1950: 351–52).

14. For women of Japanese ancestry, paid employment was common. During the first twenty years of this century, more Japanese women than women of other ethnicities worked outside the home, both on and off the plantation. Children born in Hawai'i to Japanese immigrants held dual citizenships as Japanese and Americans subjects. In the 1920s the Territorial Board of Commissioners of Education required that all public schoolteachers of Japanese descent show proof of expatriation. All had done so by 1933. Education was one of the first professions that they could enter (Tamura 1994: 38, 84).

15. See also Sinha (1995) for an exploration of sexual tensions amid colonial anxieties.

16. Senator Inouye's biography on his Web site devotes fully one-third of its prose to the details of his World War II battle experiences and his dismemberment (1996).

17. "When a black flag dropped from the fuselage to reveal the name 'Spirit of Hawaii' yesterday, Hawaii became the ninth state to have its name attached to one of the bombers. That, the generals said yesterday, was recognition of Hawaii's place in aviation and military history. Air Force aides said the name also was recognition of the congressional clout of Hawaii senior Sen. Dan Inouye, who has championed the B-2 cause since the Cold War days of the Evil Empire, when 132 were to be built at $330 million each" (Wright 1996a: A-1).

18. "Dan Medina of Moilili first saw the B-2 Spirit stealth bomber last fall and fell in love with it. 'It looks like a real spaceship,' the 6-year-old said, proudly holding up a plastic model of the bomber after viewing the real thing at Hickam Air Force Base" (Ohira 1996: A-3).

19. The price keeps going up. Robert Scheer (1997: 6) indicates that the B-2, as of fall 1997, "costs three times its weight in gold." Scheer's indignation is worth reproducing: "It is breathtaking that such an immense ripoff of the taxpayers is not bigger news, particularly at a time when vital programs for the poor are being slashed with an obnoxious glee. As this is written, 150,000 disabled children are having their Supplemental Security Income cut because their disabilities are not considered sufficiently severe. The $800 million a year saved by cutting those kids off from S.S.I. amounts to less than one-third the cost of building and maintaining a single B-2. We have already wasted an astounding $45 billion on the first twenty-one planes, and the House has authorized initial funding to build nine more at an additional cost of $27 billion" (6). Scheer wants to know, rightly, why there are no outraged editorials demanding a full-scale investigation or the appointment of a

special prosecutor, "to expose the subversives who sold us a bomber that can't go out in the rain" (7). We believe that the potent exchange relations among men in power help to answer his question.

20. The naming of *The Spirit of Hawaii* was not without controversy. About thirty protestors, objecting to the equation of Hawai'i's spirit with military might, were ejected from the ceremonies and banned from Hickam Air Force Base for one year (Wright 1996a). The *Honolulu Star-Bulletin* responded with predictable ire, scolding the protestors and alleging that they "did their cause no good by taking an anti-military stand" (1996b: A-14). Evidently still perturbed a few months later, the paper warned that "it's important that Hawaii's people show that they want the military to stay, that the anti-military activists do not represent the majority" (1996d: A-8).

It should also be noted that by 1996, when even the University of Hawai'i had adopted the Hawaiian spelling of the state's name, the neglect of the 'ōkina in naming the B-2 was itself a political act.

21. See also Heidegger (1977) and the discussion of enframing in Ferguson (1993). Our thanks to Michael Shapiro, Tom Dumm, and Bill Connolly for their help in sorting out Foucault's cryptic reflections on governmentality.

22. Hasbro Toys has produced a G.I. Joe action figure depicting a member of the 442nd Regimental Combat Team. This icon of multicultural militarism joins two African American Tuskegee airmen and a planned Colin Powell figure (*Honolulu Advertiser* 1998: A-18).

23. Many of these practices merge smoothly into everyday bureaucratic practices of discipline, classification, order, and control, practices that are not specifically militarized but that couple smoothly with military ways. For example, mandatory uniforms and/or ID tags at workplaces, sporting events, schools, and so on mark bodies and distribute them spatially.

24. We concocted this idea of the Inouye effect by appropriating and revising ideas from Graham Burchell, Colin Gordon, and Peter Miller (1991). In their book the editors invoke Gilles Deleuze's argument in *The Logic of Sense,* referring to practices of nomenclature in science, that "an effect of this kind is by no means an appearance or an illusion. It is a product which spreads or distends itself over a surface; it is strictly co-present to, and co-extensive with, its own cause, and determines this cause as an immanent cause, inseparable from its effects" (Deleuze quoted in Burchell, Gordon, and Miller 1991: ix). The editors propose the Foucault effect as "the making visible, through a particular perspective in the history of the present, of the different ways in which an activity or art called government has been made thinkable and practicable" (ix).

25. The oath of the Boy Scouts is:
On my honor I will do my best
To do my duty to God and my country

and to obey the Scout Law;
To help other people at all times;
To keep myself physically strong,
mentally awake, and morally straight.

The laws of the Boy Scouts are:

A Scout is trustworthy. A Scout tells the truth. He keeps his promises. Honesty is part of his code of conduct. People can always depend on him.

A Scout is loyal. A Scout is true to his family, friends, Scout leaders, school, nation, and world community.

A Scout is helpful. A Scout is concerned about other people. He willingly volunteers to help others without expecting payment or reward.

A Scout is friendly. A Scout is a friend to all. He is a brother to other Scouts and all the people of the world. He seeks to understand others. He respects those with ideas and customs other than his own.

A Scout is courteous. A Scout is polite to everyone regardless of age or position. He knows good manners make it easier for people to get along together.

A Scout is kind. A Scout understands there is strength in being gentle. He treats others as he wants to be treated. Without good reason, he does not harm or kill any living thing.

A Scout is obedient. A Scout follows the rules of his family, school, religion, and troop. He obeys the laws of his community and country. If he thinks those rules and laws are unfair, he tries to have them changed in an orderly manner rather than disobey them.

A Scout is cheerful. A Scout looks for the bright side of life. He cheerfully does tasks that come his way. He tries to make others happy.

A Scout is thrifty. A Scout works to pay his way and to help others. He saves for the future. He protects and conserves natural resources. He carefully uses time and property.

A Scout is brave. A Scout can face danger even if he is afraid. He has the courage to stand for what he thinks is right even if others laugh at him or threaten him.

A Scout is clean. A Scout keeps his body and mind fit and clean. He chooses the company of those who live by these same ideals. He helps keep his home and community clean.

A Scout is reverent. A Scout is reverent toward God. He is faithful in his religious duties. He respects the beliefs of others.

The oath and laws are quoted from Boy Scouts of America (1990: 550–61).

26. Roosevelt thought every boy should read *Tom Brown* and *Story of a Bad Boy*; his urging that Boy Scouts police playgrounds to keep them free of "toughs" drew on the American association of manliness with combat (Macleod 1983: 54–55).

27. These original goals were repeated in a letter to the Honorable Carl Albert, Speaker of the House of Representatives, August 3, 1976, from the National Office of the Boy Scouts of America, signed by the President and Acting Chief Scout Executive. National Office, Boy Scouts of America, 1975.

28. American Presidents since William Howard Taft have been Honorary Chairmen of the Boy Scouts of America. In the 104th Congress (1st session), 241 members had participated in scouting (Home page).

29. The biggest enrollments in traditional programs are in the 258 Boy Scout troops (6,126) and 267 Cub Scout packs (6,904) (Aloha Council 1995: 5).

30. Koichi Murakami, interviewed prior to his retirement as Scoutmaster of Troop 36 for five decades, has said that adults (Scoutmasters) should develop a boy's character, his citizenship, and his fitness but that all that boys are interested in is having fun. For Scoutmasters, "the trick is coming up with a program that accomplishes both ends." Murakami says that being a Scoutmaster has changed over the years; from being a chaperone in khaki, he is more likely to be a "stabilizing influence in a young man's life." He also made it clear to his Scouts that they could call him at home at any time to discuss a problem with him (quoted in Burlingame 1996: D-1).

31. While all the silver awards honor distinctive service to Scouts, the Silver Beaver award may be misunderstood by readers. The explanation that recipients are distinguished by their "service to youth" may still not satisfy some. The awards are given to persons (including the occasional woman) who are "active not only in Scouting, but in various religious, civic, professional and like organizations, exclusive of Scouting" (Aloha Council 1988b).

32. Schools must request JROTC units. If the enrollment of the school is less than one thousand, 10 percent of the students must express interest; if enrollment is more than one thousand, one hundred students must show interest. Schools must provide the classroom, office, storage space, phone, and photocopying access as well as paying some portions of the salaries of the instructors, who are retired military officers and noncommissioned officers. These details come from our interview with Lt. Col. Donald Barrell, U.S. Army (Ret.), Department of Education JROTC Coordinator in Hawai'i, July 10, 1995, and from Catherine Lutz and Lesley Bartlett (1995: 14). Our discussion focuses on Army JROTC practices unless otherwise noted.

33. Thirty-three hours are spent in drilling and marching; by comparison, citizenship and communication each requires around ten hours (Lutz and Bartlett 1995: 19).

34. Of the 2,740 Army JROTC instructors in 1993, seventeen were women (Lutz and Bartlett 1995: 8, 9).

35. It is estimated that the Department of Defense spends about $2 billion annually on recruitment (Rohrer 1994: 37).

36. Another recent article, "Snap and Polish," features two full-color front-page photos of JROTC male and female cadets wearing fatigues during a drill. The subhead reads: "High schoolers are using JROTC as stepping stones to discipline themselves for careers and handling leadership roles to come" (Kakesako 1996c: A-1).

37. Nationally, in 1993, 65 percent of the units were located in the South, 14 percent in the West, 12 percent in the North Central, 7 percent in the Northeast, and 2 percent overseas. Overall, "minority student participation in JROTC is approximately 54% nationwide (excluding Marine units, for which such data were not made available" (Lutz and Bartlett 1995: 6).

38. Lutz and Bartlett (1995) point to the RedCat Career Academy, a naval JROTC unit, as a possible exception since its entrance criteria include students with high absentee rates, low motivation, economic disadvantages, and poor academic performance (38, n. 36).

39. Enrollment in all services' public school programs was 2,066 with a further 754 in private schools (Hawaii Department of Education 1995a).

40. There is no question that the JROTC program in Hawai'i is committed to serving the youths enrolled; its former coordinator, Lt. Col. Barrell, was a very successful instructor for a few years at a school where there can be notable antagonisms to both the military and Caucasians. The current coordinator is Barrell's former coworker, Lt. Col. Jeff Tom from Hawai'i. The program instructors invest much of themselves and their time in assisting and counseling their charges. All quotations by Col. Barrell are from our interview with him (Barrell 1995).

41. The definition of sports in Hawai'i's high schools reflects the cozy relation of the military to education. Marksmanship is defined by the state as a sport. Schools with JROTC programs and those without both have received weapons and transportation from the military to start up their extracurricular activities involving air rifles and .22 caliber rifles. No doubt, financially strapped schools are grateful for this assistance, yet it ties the schools more tightly to military priorities. Our thanks to Neal Takamuri, athletic director at McKinley High School in Honolulu, for useful background information on this topic.

42. "A very unusual proceeding on the part of the Honolulu Chamber," commented Admiral Samuel W. Very, Commandant of the Honolulu Naval Station (Hodge and Ferris 1950: 26).

43. It is estimated by some that corporate taxes are low, but this complex calculation is not readily available. The state of Hawai'i collected more than $4 billion in revenues in 1994. The single largest source of that came from the general excise tax (GET). Unlike a sales tax, the GET is a pyramided tax, that is, it is collected at "every level of production, distribution, and sale." It is estimated that Hawai'i would need a sales tax rate of 18 to

20 percent to equal the current 4 percent GET rate (Frierson and Thorndike 1996: 9). According to the 1995 *State of Hawaii Data Book*, while collecting $1,287,225,000 in general excise taxes, the state in 1992 took in $889,532,000 from individual income taxes and only $37,514,000 in corporate taxes (Hawaii Department of Business 1995: 257). Using Hawaii Department of Business, Economic Development, and Tourism data, the Bank of Hawaii showed Hawai'i's maximum corporate income tax rate to be 6.4 percent. Iowa leads the states with a corporate income tax of 12 percent (Bank of Hawaii 1998: 24).

44. Opponents were dubious inasmuch as the battleship had just undergone an overhaul and because many of the skilled workers needed for other work would accompany the group. Additionally, housing for the dependents would have created an even greater shortage for O'ahu residents.

45. A convenient amnesia seems to have shielded them from recalling a history of armed intervention in the affairs of the kingdom. See chapters 1 and 2.

46. Dennis Fujii, chief of Community Relations at Hickam Air Force Base, observed a gradual positive change during the 1980s in the attitude of local communities toward the military. He credited the Military Affairs Council for stimulating recognition "that the military is an industry, and it plays a significant financial part in Hawai'i (interview with Fuji 1991).

47. The Bank of Hawaii estimates that the multiplier effect of each tourist dollar spent in Hawai'i is 2.04, that is, that enough secondary expenditures are generated by the original dollar to total $2.04. This figure is lower than that for military spending because of what the bank's report refers to as "leakage," the export of half of each dollar sent out of state to hotel owners, banks, foreign airlines, and other suppliers of goods and services (1984: 2). According to the 1995 *State of Hawaii Data Book*, visitor-related expenditures in Hawai'i amounted to $11,035,700,000 in 1994 and was responsible for 245,400 jobs (Hawaii Department of Business 1995: 208).

48. Hawai'i is linguistically unfamiliar as well. Lamenting that military personnel "think *mahalo* is Hawaiian for *trash* because it's on every garbage can," Director of Family Support Services Christine Burnette noted that the Air Force briefs newcomers in an effort "to give people a little bit of cultural sensitivity" (interview with Burnette 1991).

49. The financial difficulties of enlisted and junior ranks mirror the larger situation of low- and moderate-income people in Hawai'i. Reflecting a national trend, the gap between rich and poor in Hawai'i is growing; in 1991 Hawai'i ranked "32nd on the list of the 50 states in income inequality among its residents" (First Hawaiian Bank 1991: 1).

50. Federal impact aid is a form of revenue directed to states in which there is significant federal activity. It recognizes that the presence of federal

activity incurs for local governments expenditures that are often unrecoverable through the usual education revenue sources: property, income, and/or sales taxes. In fiscal 1990–91, impact aid for the 36,152 federally connected children amounted to 11 percent of the $192,060,392 it cost the state to educate them. "Federally connected children" include more than military offspring. In this study, military and nonmilitary children were lumped together in calculating the amount of state funds spent to educate military family members (Lee 1993: 34).

51. The *Advertiser,* however, lacks the tireless enthusiasm for the military presence evinced in *Star-Bulletin* op-ed pieces by its contributing editor, Bud Smyser. A random selection of those titles reads: "Let Military Resume Bombing of Kahoolawe," "How Hawaii Benefited from World War II," "Reconciling Community, Military Land Needs," "Military Commands Are an Asset for Hawaii," "Armed Forces Are Still Important to Hawaii," "Pacific Fleet Shouldn't Be Cut Further."

52. Many of the Hawaiian language newspapers of the nineteenth century have not yet been used in historical research, because most historians of Hawai'i have not been able to read the Hawaiian language. Their interpretations, lacking these oppositional voices, have been significant in shaping the hegemonic voice. Younger Hawaiian scholars, literate in their language, are writing different histories. For example, see Silva (1997 and forthcoming), Kame'eleihiwa (1992), and Patrinos (1995).

53. Chapin (1996: 316) writes that some observers suggest that this gives the Gannett Corporation ownership of both papers since it also owns the Hawaii News Agency, the joint business and production entity.

54. These statistics are cited from an August 6, 1991, letter from Derek Davies, former editor of the *Far Eastern Economic Review,* then at the East-West Center in Honolulu as an editor in residence, to Fletcher Knebel, chair of Honolulu Community-Media Council.

55. Using funds available as peace dividends, the military is attempting to alter its ways, making commitments to the prevention of future environmental degradation and the repair of some past disasters. Tummons reports a project of the Marine Corps to improve habitat for water and sea birds that nest around their Kāne'ohe base (1994: 10). Virtually all services now have base environmental programs. They contract with environmental organizations such as the Nature Conservancy and the Environmental Center at the University of Hawai'i. According to John Harrison, director of the center, the military is making great efforts to develop its own capacities for sound environmental management. Our thanks to John Harrison for this information.

56. The "ceded lands" are those that belong to Native Hawaiians but that were casually lumped together with government lands when the United

States annexed Hawai'i. Under the terms of the act creating Hawai'i as a state, a portion of the income from those lands is to be used for the benefit of Native Hawaiians.

57. The publication's title and the cover photo on the October/November 1994 issue offer a bewildering yet suggestive semiotics that evokes the liminality of the homosocial/homosexual boundary (Sedgwick in Dumm 1994: 52–53): two navy men stand close together on the deck of a ship; the one who is clothed assists another, wearing a diving brief, to don his scuba gear by steadying him with a braced arm and handing him the rest of his equipment. The heady mixture of nudity, proximity, and "guy stuff" (Barry 1997) is inescapably erotic.

Bibliography

Adamski, Mary. 1997. "Niihau Residents Have Concerns," *Honolulu Star-Bulletin*: A-3. June 24.

Albertini, Jim, Nelson Foster, Wally Inglis, and Gil Roeder. 1980. *The Dark Side of Paradise: Hawaii in a Nuclear World*. Honolulu: Catholic Action of Hawaii/Peace Education Project.

Alcalay, Glenn H. 1984. "Maelstrom in the Marshall Islands: The Social Impact of Nuclear Weapons Testing," in *Micronesia as a Strategic Colony: The Impact of U.S. Policy on Micronesian Health and Culture*, ed. Catherine Lutz. Cambridge, Mass.: Cultural Survival, 25–36.

Alexander, William J. 1984. "Missiles, the Military, and the Marshalls," in *Micronesia as a Strategic Colony: The Impact of U.S. Policy on Micronesian Health and Culture*, ed. Catherine Lutz. Cambridge, Mass.: Cultural Survival, 7–24.

Ali, Mehmed, 1991. "*Loko I'a* and Modern Aquaculture." Unpublished paper, Department of Political Science, University of Hawai'i, Honolulu.

Ali, Mehmed, Kathy E. Ferguson, and Phyllis Turnbull (with Joelle Mulford). 1991. "Gender, Land and Power: Reading the Military in Hawai'i," presented at the fifteenth World Congress of the International Political Science Association, Buenos Aires, July 21–25.

Allen, Gwenfread. 1950. *Hawaii's War Years 1941–1945*. Honolulu: University of Hawaii Press.

Aloha Council, Boy Scouts of America. 1988a. *Hawaii's Distinguished Citizen*. Program for dinner honoring Senator Daniel K. Inouye, Sheraton Waikīkī Ballroom, Honolulu, Hawai'i. August 17.

———. 1988b. *Annual Report*. Honolulu.

———. 1995. *Annual Report*. Honolulu.

Althusser, Louis. 1971. *Lenin and Philosophy and Other Essays*, trans. Ben Brewster. New York: Monthly Review Press.

Altonn, Helen. 1992. "No Leukemia Link Found to Lualualei," *Honolulu Advertiser*: A-1, A-8. July 24.

Ambrose, Greg. 1997. "Say Y.E.S.," *Honolulu Star-Bulletin*: B-1. July 28.

American Friends Service Committee Hawai'i (AFSC). n.d. "Caravan for Peace," Church of the Crossroads, Honolulu, Hawai'i.

———. N.d. *Hawaiian Lands and the Military,* Hawai'i Area Program Office, Honolulu, Hawai'i.

Andrews, Edmund L. 1997. "Europeans Take Boeing to Brink," *New York Times*: A-1, C-5. July 24.

Anthony, J. Garner. 1975. *Hawaii under Army Rule*. Honolulu: University of Hawaii Press.

Avineri, Shlomo. 1972. *Hegel's Theory of the Modern State*. London: Cambridge University Press.

Bailey, Beth, and David Farber. 1992. *The First Strange Place: Race and Sex in World War II Hawaii*. Baltimore: Johns Hopkins University Press.

Bakhtin, M. M. 1981. *The Dialogic Imagination,* ed. Michael Holquist and trans. Caryl Emerson and Michael Holquist. Austin: University of Texas Press.

Bakutis, Bunky. 1977. "'Bomb 'Ohana' T-Shirts No Joke to Hawaiians," *Honolulu Advertiser*: A-3. March 31.

Bank of Hawaii. 1984. *Business Trends*. Honolulu, July/August.

———. 1997. *Hawaii 1997: Annual Economic Report*. Vol 46. Honolulu.

———. 1998. *Business Trends*. Honolulu, July/August.

Barrett, Greg. 1996. "Military Ouster Sought: War-Protest Group Rallies at Bases on Memorial Day," *Honolulu Advertiser*: A-3. May 28.

Barry, Dave. 1997. "Rubber-Band Flight Suit Another 'Guy' Thing," *Honolulu Advertiser*: E-2. August 3.

Barthes, Roland. 1978. *Mythologies,* New York: Hill and Wang.

———. 1991. "Right in the Eyes," in *The Responsibility of Forms: Critical Essays on Music, Art and Representation,* trans. Richard Howard. Berkeley and Los Angeles: University of California Press, 237–42.

Baudrillard, Jean. 1981. *For a Critique of the Political Economy of the Sign*. St. Louis: Telos Press.

Bell, Roger. 1984. *Last among Equals: Hawaiian Statehood and American Politics*. Honolulu: University of Hawaii Press.

Berger, John. 1979. *Ways of Seeing*. Harmondsworth, England: Penguin Books.

Berlant, Lauren. 1993. "The Theory of Infantile Citizenship," *Public Culture* 5: 395–410.

Bhabha, Homi K. 1990. "DissemiNation: Time, Narrative, and the Margins of the Modern Nation," in *Nation and Narration,* ed. Homi K. Bhabha. New York: Routledge, 291–322.

Billig, Priscilla Pérez. 1997. "Is Hawaii Prepared for an Oil Spill?" *Mālamalama* 21 (1): 16–17.

Bingham, Hiram. 1981. *Residence of Twenty-One Years in the Sandwich Islands; Or, the Civil, Religious, and Political History of Those Islands.* Reprinted. Rutland, Vt.: Charles E. Tuttle.

Bishop, S. E. 1888. "Why Are Hawaiians Dying Out? Or, Elements of Disability for Survival among the Hawaiian People." Presented to the Honolulu Social Science Association.

Blount, James H. 1893. *Hawaiian Islands.* Report to President Cleveland from the Special Commissioner of the United States. U.S. Congress, House. Executive Document No. 47. 53 Cong. 2 Session. Washington, D.C.

Bone, Robert W. 1996. "Majuro and Bikini," *Honolulu Advertiser:* F-1, F-2. November 24.

Borg, Jim. 1990. "Hawaiians Want Lualualei," *Honolulu Advertiser:* C-9. August 3.

Borreca, Richard. 1992. "Inouye Says Life Has Been 'Living Hell,'" *Honolulu Star-Bulletin:* A-3. December 16.

Botticelli, Ann. 1995. "Inouye to Clinton: Forget Apology," *Honolulu Advertiser:* A-1. January 15.

Boyer, Peter J. 1996. "A Reporter at Large: Admiral Boorda's World," *New Yorker:* 68–86. September 16.

Boylan, Dan. 1996. "Courting Supreme Controversy," *Midweek:* A-10, A-11. August 21.

Boy Scouts of America. 1990. *The Boy Scout Handbook,* 10th ed. Irving, Tex.: Boy Scouts of America.

———. Home page. http://www.bsa.scouting.org/

Brotherston, Gordon. 1992. *Book of the Fourth World: Reading the Native Americas through Their Literature.* London: Cambridge University Press.

Brown, Hubert E. 1945. "The Effects of Closing Houses of Prostitution," presented at the Social Protection Committee of the Honolulu Council of Social Agencies. March 7.

Bryant, C. D. 1976. *Friendly Fire.* New York: Putnam.

Buck, Elizabeth. 1993. *Paradise Remade: The Politics of Culture and History in Hawai'i.* Philadelphia: Temple University Press.

Burchell, Graham, Colin Gordon, and Peter Miller, eds. 1991. *The Foucault Effect: Studies in Governmentality.* Chicago: University of Chicago Press.

Burlingame, Burl. 1990. "New Life, New Look for Army Museum." *Honolulu Star-Bulletin:* B-1, B-3. August 27.

———. 1996. "Scoutmaster 'Kobu' Murakami: An Inspiration for More Than 50 Years," *Honolulu Star-Bulletin*: D-1. July 23.

Burris, Jerry. 1988. "Go for Broke," *Honolulu Advertiser*: A-1, A-4. November 27.

Bushnell, Andrew. 1992. "The 'Horror' Reconsidered: An Evaluation of the Historical Evidence for Population Decline in Hawai'i, 1778–1803." Unpublished paper, Department of History, University of Hawai'i.

Bushnell, O. A. 1993. *The Gifts of Civilization: Germs and Genocide in Hawai'i.* Honolulu: University of Hawai'i Press.

Cabacungan, Kathryn. 1994. "Punchbowl Holy Ghost Fiesta Was a Special Day," *Honolulu Advertiser*: G-12. May 1.

Cahill, Robert S. 1975. "Testimony to the Board of Regents." Unpublished manuscript, Department of Political Science, University of Hawai'i. April 23.

Campbell, David. 1992. *Writing Security: United States Foreign Policy and the Politics of Identity.* Minneapolis: University of Minnesota Press. [Revised edition, 1998.]

Carlson, Douglas. 1990. *Punchbowl: The National Memorial Cemetery of the Pacific.* Aiea, Hawai'i: Island Heritage Publishing.

Carter, Paul. 1987. *The Road to Botany Bay: An Exploration of Landscape and History.* Chicago: University of Chicago Press.

Castanha, Anthony. 1996. "Roles of Non-Hawaiians in the Hawaiian Sovereignty Movement." Master's thesis, Political Science Department, University of Hawai'i.

Chadwin, Dean. 1996. "Ohanaball," *Honolulu Weekly* 6 (48): 5–8. November 27–December 3.

Chamber of Commerce of Hawaii. N.d. *Military Affairs Council.* Pamphlet.

———. 1992. *Stand Tall, Stand Proud!* Pamphlet for Hawaii Military Week. May 11–16.

———. 1996. *Military Affairs Council.* Pamphlet.

Chapin, Helen Geracimos. 1996. *Shaping History: The Role of Newspapers in Hawai'i.* Honolulu: University of Hawai'i Press.

Chapman, Don. 1995. "Regal Bearings: The Aloha Festival's Royal Court Traditions Hail from the Monarchy," *Hawaii Magazine*: 46–51. October.

Christic Institute. 1987. "A Christic Institute Special Report: The Contra-Drug Connection." Washington, D.C.: Christic Institute.

Chuzo, Kazuo. 1984. *Hiroshima Maidens: The Nuclear Holocaust Retold,* trans. Asahi Evening News. Tokyo: Asahi Shimbum.

Cixous, Hélène, and Catherine Clément. 1986. *The Newly Born Woman,* trans. Betsy Wing. Minneapolis: University of Minnesota Press.

Clark, Roger S. 1980. "Self-Determination vs Free Association—Should the

United Nations Terminate the Pacific Islands Trust?" *Harvard International Law Journal* 21 (1): 1–86.

Clark, Roger S., and Sue Roff. 1987. *Micronesia, the Problem of Palau.* Report #63. London: Minority Rights Project.

Coffman, Tom. 1973. *Catch a Wave: A Case Study of Hawaii's New Politics.* Honolulu: University of Hawaii Press.

Cohn, Carol. 1987. "Sex and Death in the Rational World of Defense Intellectuals," *Signs: Journal of Women in Culture and Society* 12 (Summer): 687–718.

Connolly, William E. 1991. *Identity/Difference: Democratic Negotiations of Political Paradox.* Ithaca, N.Y.: Cornell University Press.

———. 1995. *The Ethos of Pluralization.* Minneapolis: University of Minnesota Press.

Cook, James. 1961. *The Journals of Captain James Cook on His Voyages of Discovery,* ed. J. C. Beaglehole. 3 vols. Cambridge: Cambridge University Press. Vol. 2: *The Voyage of the Resolution and Adventure 1772–1775.*

———. 1967. *The Journals of Captain James Cook,* ed. J. C. Beaglehole. Cambridge: Cambridge University Press. Vol. 3, parts 1, 2: *The Voyage of the Resolution and Discovery.*

Cooper, George, and Gavan Daws. 1985. *Land and Power in Hawaii: The Democratic Years.* Honolulu: Benchmark Press.

Council for a Livable World Education Fund. 1995. "B-2 Bomber Factsheet."

Culler, Jonathan. 1988. *Framing the Sign.* Norman: University of Oklahoma Press.

Dallmayr, Fred. 1993. *G. W. F. Hegel: Modernity and Politics.* Newbury Park, Calif.: Sage Publications.

David, Andrew, ed. 1988. *The Charts and Coastal Views of Captain Cook's Voyages.* Vol. 1, *The Voyages of the* Endeavour *1768–1771.* London: Hakluyt Society Publications.

Davis, Lynn Ann, and Nelson Foster. 1988. *A Photographer in the Kingdom: Christian J. Hedemann's Early Images of Hawai'i.* Bishop Museum Special Publication 85. Honolulu: Bishop Museum Press.

Davis, Stu. 1996. "Playing with Fire," *Honolulu Weekly* 6 (29): 8–9. July 17.

Daws, Gavan. 1968. *The Shoal of Time: A History of the Hawaiian Islands.* New York: Macmillan.

———. 1980. *A Dream of Islands: Voyages of Self-Discovery in the South Seas.* New York: W. W. Norton.

DeCambra, Ho'oipo. 1997. "Pride in Living by Moving towards Self-Reliance and Self-Sufficiency through the Creation of Projects Directly Benefitting Poor Families," presented at the Women's Brown Bag Seminar, Office of Women's Research, University of Hawai'i. January 24.

Delgado, James P. 1996. *Ghost Fleet: The Sunken Ships of Bikini Atoll.* Honolulu: University of Hawai'i Press.

de Man, Paul. 1984. "The Epistemology of Metaphor," in *Language and Politics,* ed. Michael J. Shapiro. New York: New York University Press, 195–214.

Diamond, John. 1997. "Why Consolidation Has Become the Rule in Defense Industries," *Honolulu Advertiser*: B-8, B-7. July 4.

Dibble, Sheldon. 1839. *History and General Views of the Sandwich Islands' Mission.* New York: Taylor and Dodd.

Diller, Elizabeth and Ricardo Scofidio, eds. 1994. *Visite aux armees: Tourismes de guerre* (Back to the front: Tourisms of war). France: F.R.A.C. Basse-Normandie.

Dillon, Michael. 1995. "Sovereignty and Governmentality: From the Problematics of the 'New World Order' to the Ethical Problematic of the World Order," *Alternatives* 20: 323–68.

Dingeman, Robbie. 1996. "Inouye's Power of the Pork," *Honolulu Star-Bulletin*: A-1. March 17.

DiPalma, Carolyn. 1995. "Elision and Specificity Written as the Body: Sex, Gender, Race, Ethnicity in Feminist Theory." Ph.D. diss., Department of Political Science, University of Hawai'i.

Doty, Roxanne Lynn. 1996a. "The Double-Writing of Statecraft: Exploring State Responses to Illegal Immigration," *Alternatives* 21: 171–89.

———. 1996b. *Imperial Encounters: The Politics of Representation in North-South Relations.* Minneapolis: University of Minnesota Press.

Dower, John. 1986. *War without Mercy: Race and Power in the Pacific War.* New York: Pantheon.

Draper, Theodore. 1997. "Is the CIA Necessary?" *New York Review of Books* 44 (13): 18–22.

Dumm, Thomas L. 1994. *united states.* Ithaca, N.Y.: Cornell University Press.

———. 1996. *Michel Foucault and the Politics of Freedom.* London: Sage Publications.

Ebert, Teresa L. 1991. "Political Semiosis in/or American Cultural Studies," *American Journal of Semiotics* 8 (1–2): 113–35.

Ehrenhaus, Peter. 1989. "Commemorating the Unwon War: On *Not* Remembering Vietnam," *Journal of Communication* 39 (1): 96–107.

Elshtain, Jean Bethke. 1987. *Women and War.* New York: Basic Books.

Enloe, Cynthia. 1990. *Bananas, Beaches and Bases: Making Feminist Sense of International Politics.* Berkeley and Los Angeles: University of California Press.

———. 1993. *The Morning After: Sexual Politics at the End of the Cold War.* Berkeley and Los Angeles: University of California Press.

Ensign, Tod. 1992. "Military Resisters during Operation Desert Shield/Storm," in *Collateral Damage: The New World Order at Home and Abroad,* ed. Cynthia Peters. Boston: South End Press, 279–97.

Farrell, B. H. 1982. *Hawaii: The Legend That Sells.* Honolulu: University of Hawaii Press.

Federation of American Scientists. 1996. "Stop the B-2 Bomber." http://www.fas.org/pub/gen/mswg/stealth/

Ferguson, Kathy E. 1993. *The Man Question: Visions of Subjectivity in Feminist Theory.* Berkeley and Los Angeles: University of California Press.

Ferrar, Derek, and Julie Steele. 1993. "100 Years of Subjugation," *Honolulu Weekly* 3 (2): 4–8. January 13.

First Hawaiian Bank. 1991. *Economic Indicators.* Honolulu. January/February.

———. 1993a. *Economic Indicators.* Honolulu. March/April.

———. 1993b. *The Role of the Military in Hawaii's Economy.* Honolulu. May.

———. 1995. *Economic Indicators,* supplement. Honolulu. May/June.

Foodbank Review. 1996. 21. September.

Foucault, Michel. 1977. "Nietzsche, Genealogy, History," in *Language, Countermemory, Practice: Selected Essays and Interviews,* ed. Donald F. Bouchard. Ithaca, N.Y.: Cornell University Press, 139–54.

———. 1978. *The History of Sexuality.* Vol. 1. An Introduction. Trans. Robert Hurley. New York: Pantheon.

———. 1979. *Discipline and Punish: The Birth of the Prison,* trans. Alan Sheridan. New York: Vintage Books.

———. 1988a. "The Art of Telling the Truth," in *Michel Foucault: Politics, Philosophy, Culture,* ed. Lawrence Kritzman, trans. Alan Sheridan et al. New York: Routledge, 86–95.

———. 1988b. "Critical Theory/Intellectual History," in *Michel Foucault: Politics, Philosophy, Culture,* ed. Lawrence Kritzman, and trans. Alan Sheridan et al. New York: Routledge, 17–46.

———. 1991a. "Politics and the Study of Discourse," in *The Foucault Effect: Studies in Governmentality,* ed. Graham Burchell, Colin Gordon, and Peter Miller. Chicago: University of Chicago Press, 53–72.

———. 1991b. "Governmentality," in *The Foucault Effect: Studies in Governmentality,* ed. Graham Burchell, Colin Gordon, and Peter Miller. Chicago: University of Chicago Press, 87–104.

Frankenberg, Ruth. 1993. *White Women, Race Matters: The Social Construction of Whiteness.* Minneapolis: University of Minnesota Press.

Frear, Walter F. 1935. "Anti-Missionary Criticism: With Reference to Hawaii." Presented to the Honolulu Social Science Association.

Frierson, Jill, and Suzanne Thorndike, eds. 1996. *The Little Red Budget*

Book: How the State of Hawaii Collects, Budgets and Spends Your Tax Dollars. Prepared by the research arm of the Hawaii State Senate Republican Caucus. Honolulu: Little Red Budget Book Project.

Fuchs, Lawrence H. 1961. *Hawaii Pono: A Social History.* New York: Harcourt, Brace, and World.

Glauberman, Stu. 1995. "Tour Enlists Military Sites," *Honolulu Advertiser*: D-1, D-2. April 3.

Goodyear-Ka'ōpua, Jennifer Noelani. 1997. "The Genealogy of French Nuclear Colonization in the Pacific," *Student Working Papers Series*. Vol. 2, ed. Louise Kubo. Honolulu: Office for Women's Research, University of Hawai'i, 4–13.

Gordon, Colin. 1991. "Governmental Rationality: An Introduction," in *The Foucault Effect: Studies of Governmentality*, ed. Graham Burchell, Colin Gordon, and Peter Miller. Chicago: University of Chicago Press, 1–51.

Goss, Jon. 1993. "Placing the Market and Marketing the Place: Tourist Advertising of the Hawaiian Islands, 1972–92," *Environment and Planning D: Society and Space* 11: 663–88.

Gray, Francine de Plessix. 1972. *The Sugar-Coated Fortress.* New York: Random House.

Gregory, Eric. 1995. "Hickam Memorial Honors War Vets," *Honolulu Advertiser*: A-3. August 24.

Griffin, John. 1991. "Uncertainty, Instability: New Pacific Watchwords," *Honolulu Star-Bulletin and Advertiser*: F-3. March 31.

———. 1997. "Pearl Harbor to Become an Even Bigger Visitor Attraction," *Honolulu Advertiser*: B-1, B-4. September 21.

Griffin, Susan. 1992. *A Chorus of Stones: The Private Life of War.* New York: Doubleday Anchor Books.

Grimshaw, Patricia. 1989. *Paths of Duty: American Missionary Wives in Nineteenth-Century Hawaii.* Honolulu: University of Hawaii Press.

Griswold, Charles. 1986. "The Vietnam Veterans Memorial and the Washington Mall: Philosophical Thoughts on Political Iconography," *Critical Inquiry* 12 (Summer): 688–719.

Gross, Jane. 1992. "Accusations against Hawaii Senator Meet a Silence in His Seat of Power," *New York Times*: C-10. December 14.

Grosz, Elizabeth. 1995. *Space, Time, and Perversion.* New York: Routledge.

Hall, Dana Naone, ed. 1985. *Malama: Hawaiian Land and Water.* Honolulu: Bamboo Ridge Press.

Haraway, Donna. 1989. *Primate Visions: Gender, Race, and Nature in the World of Modern Science.* New York: Routledge.

———. 1997. *Modest Witness @ Second Millennium.* New York: Routledge.

Hartung, William D. 1997. "Military Monopoly," *Nation* 264 (2): 6–7. January 13/20.

Hartwell, Jay. 1985. "Battle over Koas of Kilauea," *Honolulu Advertiser*: A-3. March 4.

Hau'ofa, Epeli. 1994. "Our Sea of Islands," *Contemporary Pacific* 6 (1): 147–61.

Hawaiian Phrase Book. 1968. Rutland, Vt., and Tokyo: Charles E. Tuttle. First published in 1906.

Hawaiian Sugar Planters' Association. 1921. *The Sugar Industry of Hawaii and the Labor Shortage: What It Means to the United States and Hawaii.* Honolulu.

Hawaii Department of Business, Economic Development and Tourism (DBED), Research and Economic Analysis Division, Statistics Branch. 1994. *State of Hawaii Data Book: A Statistical Abstract.* Honolulu.

———. 1995. *State of Hawaii Data Book: A Statistical Abstract.* Honolulu.

Hawaii Department of Education. 1995a. *Memorandum on JROTC Enrollment for the School Year 1994–1995.*

———. 1995b. JROTC Coordination Office. "Backgrounder #1." April 17.

Hawaii Department of Human Services. 1994. *Child Abuse and Neglect in Hawaii.* Honolulu.

Hawaii Economic Development Project, Homeport Task Force. 1984. *Hawaii's Program for Supporting the Navy's Homeporting Plan.* Chapter 3, part 1.

Hawai'i Ecumenical Coalition. 1995. "Is It Time to Give Back Hawai'i?" *New York Times*: A-6. July 7.

Hawai'i Monitor. 1992. "Very Few Political Leaders Respond Well to Inouye Harassment Charges," 4. November.

Hawaii Revised Statutes. 1993. Annotated. Revisor of Statutes. Honolulu, Hawaii.

Hegel, Georg Wilhelm Friedrich. 1967. *The Phenomenology of Mind*, trans. J. B. Baille. New York: Harper and Row.

———. 1975. *Lectures on the Philosophy of World History*, trans. H. B. Nisbet. Cambridge: Cambridge University Press.

Heidegger, Martin. 1977. *The Question Concerning Technology and Other Essays*, trans. William Lovitt. New York: Harper Colophon Books.

Herman R. Douglas K. 1995. "Kālai'āina—Carving the Land: Geography, Desire and Possession in the Hawaiian Islands." Ph.D. diss., Department of Geography, University of Hawai'i.

———. 1996. "The Dread Taboo: Human Sacrifice, and Pearl Harbor," *Contemporary Pacific* 8(1): 81–125.

Hibbard, Benjamin H. 1965. *A History of the Public Land Policies.* Madison: University of Wisconsin Press.

Hinsdale, B. A. 1988. *The Old Northwest: With a View of the Thirteen*

Colonies as Constituted by the Royal Charters. New York: Townsend MacCoun.

Hirschmann, Nancy J. 1992. *Rethinking Obligation: A Feminist Method for Political Theory.* Ithaca, N.Y.: Cornell University Press.

Hobbes, Thomas. 1968. *Leviathan,* ed. C. B. MacPherson. Middlesex, England: Penguin Books.

Hodge, Clarence L., and Peggy Ferris. 1950. *Building Honolulu: A Century of Community Service.* Honolulu: Chamber of Commerce of Honolulu.

Honolulu Advertiser. 1946. July 4: 5.

———. 1990a. "Johnston Island Facility Gets the Nod from Inouye," August 15: A-1, A-4.

———. 1990b. Letter to the editor, November 10: A-15.

———. 1992. "Inouye Praises Citizen-Patriots," April 27: A-6.

———. 1993. "Military Gets Big Funding," November 12: A-2.

———. 1994. "State and Military to Salute Each Other in Week-Long Festival," May 15: A-7.

———. 1995. "ROTC Cadets Dig In to the Call of Dirty Duty," April 2: A-1, A-8.

———. 1996b. "Foodbank Benefit a Tribute to Inouye," June 18: B-3.

———. 1996c. "Sailor Attacked at Waimea Bay," June 20: A-3.

———. 1996d. "CIA and Cocaine," August 26: A-8.

———. 1998. "GI Joe Doll to Honor 442nd Soldiers," April 12: A-18.

Honolulu Star-Bulletin. 1934. March 10: 7.

———. 1946. "Cmdr. Gardner Wins Acquittal in Damon Tract 'Riot' Court Martial," April 23: 2.

———. 1948. March 23: 5.

———. 1985. Letter to the editor, July 11: 12.

———. 1991. March 12: A-17.

———. 1993. "Abercrombie's Proposal," editorial, October 30: A-6.

———. 1995a. "Kauai's Economy Will Get Boost from Range," editorial, August 1: A-10.

———. 1995b. "Sen. Inouye Honored by Flyover in Mississippi," June 19: A-8.

———. 1996a. Northrop Grumman advertisement, May 24: A-12.

———. 1996b. "Bomber Appropriately Named for Hawaii," editorial, May 29: A-14.

———. 1996c. "Inouye and Dole Go Back a Long Way," June 17: A-3.

———. 1996d. "Importance of Military Spending in Hawaii," editorial, July 8: A-8.

———. 1996e. "U.S. Jets Kill Iraq Radar as MiGs Answer Raiders," September 4: A-1, A-6.

———. 1997. "Hawaii May Receive Millions in Federal Money for Military Projects," July 11: A-5.

Honolulu Star-Bulletin and Advertiser. 1990. "The '90s: 'Decade of Great Crises?'" March 18: B-1, B-3.

Horowitz, Adam. 1991. "Home on the Range." Adam Horwitz/Equitorial Films.

Howes, David, ed. 1991. *The Varieties of Sensory Experience: A Sourcebook in the Anthropology of the Senses.* Toronto: University of Toronto Press.

Iles, William P. 1978. "Quest of Blame: Inquiries Conducted 1941–1946 into America's Involvement in the Pacific War." Ph.D. diss., University of Iowa.

"Important Views on the Question of Annexation," December 30, 1897, January 5, 1898, and January 15, 1898. Reprinted from *The Evening Star.* Honolulu.

Inglis, Ken. 1989. "Men, Women, and War Memorials: Anzac Australia," in *Learning about Women: Gender, Politics and Power,* ed. Jill Conway, Susan C. Bourque, and Joan W. Scott. Ann Arbor: University of Michigan Press, 35–60.

Inouye, Daniel K. 1975. "Daniel K. Inouye," John A. Burns Oral History Project, University of Hawaii, Honolulu. August 6.

———. 1996. Home page. http://www.senate.gov/~inouye/bio.html

Iriye, Akira. 1972. *Pacific Estrangement: Japanese and American Expansion, 1897–1911.* Cambridge, Mass.: Harvard University Press.

Ishtar, Zohl de. 1994. *Daughters of the Pacific.* North Melbourne, Australia: Spinifex Press.

Ivins, Molly. 1996. "Subsidizing the 'Merchants of Death,'" *Honolulu Star-Bulletin:* A-8. August 5.

Jackson, Kenneth T., and Camilo Jose Vergara. 1989. *Silent Cities: The Evolution of the American Cemetery.* New York: Princeton Architectural Press.

Japan at War. 1980. Alexandria, Va.: Time-Life Books.

Jay, Martin. 1988. "Scopic Regimes of Modernity," in *Visions and Visuality,* ed. H. Foster. Seattle: Bay Press.

———. 1994. *Downcast Eyes: The Denigration of Vision in Twentieth Century French Thought.* Berkeley and Los Angeles: University of California Press.

Jenkins, Robert S. 1997. "Hawaii's New Reciprocal Beneficiary Law," Pride Insurance and Financial Services, Honolulu.

Joesting, Edward. 1972. *Hawaii: An Uncommon History.* New York: Norton and Colm.

Johnson, Hildegaard Binder. 1976. *Order upon the Land: The U.S. Rectangular Survey and the Upper Mississippi Country.* London: Oxford University Press.

Jones, Kathleen B. 1993. *Compassionate Authority: Democracy and the Representation of Women*. New York: Routledge.

Judd, Lawrence M. 1971. *Lawrence M. Judd and Hawaii: An Autobiography*, with Hugh W. Lytle. Rutland, Vt.: Charles Tuttle.

Ka'ai, Sam. 1987. *Does the Mo'o Live Here Anymore? Commentaries on Hawaiian Thought*. Keynote address presented to Third International Conference on Thinking, Waikīkī, Honolulu, January 4.

Kakesako, Gregg K. 1993. "Battlefield Act Haunts Sen. Inouye," *Honolulu Star-Bulletin*: A-1, A-8. March 27.

———. 1996a. "Navy Wives Endure Their Own Waiting Game," *Honolulu Star-Bulletin*: A-4. September 5.

———. 1996b. "Honor Overdue: Asian-American Soldiers Are Getting a Second Chance at Medal of Honor," *Honolulu Star-Bulletin*: A-1, A-6. November 11.

———. 1996c. "Snap and Polish," *Honolulu Star-Bulletin*: A-1, A-10. December 10.

———. 1997. "It Seemed Like the Whole Harbor Had Blown Up," *Honolulu Star-Bulletin*: A-3. December 5.

Kamakau, Samuel Manaia Kalani. 1964. *Ka po'e kahiko: The People of Old*, trans. Mary Kawena Pukui, ed. Dorothy B. Barrere. Bernice P. Bishop Museum Special Publication no. 51. Honolulu: Bishop Museum Press.

———. 1976. *The Works of the People of Old*, trans. Mary Kawena Pukui, ed. Dorothy B. Barrere. Bernice P. Bishop Museum Special Publication no. 61. Honolulu: Bishop Museum Press.

———. 1991. *Tales and Traditions of the People of Old*. Honolulu: Bishop Museum Press.

———. 1992. *The Ruling Chiefs of Hawaii*. Rev. ed. Honolulu: Kamehameha Schools Press.

Ka Mana A Ka 'Āina. 1989. "Stop the Misery from Home-Porting in Peril Harbor!" *Bulletin of the Pro-Hawaiian Sovereignty Working Group* 1(4). December.

Kame'eleihiwa, Lilikalā. 1992. *Native Land and Foreign Desires: Pehea Lā E Pono Ai?* Honolulu: Bishop Museum Press.

Kanda, Mikio, ed. 1989. *Widows of Hiroshima: The Life Stories of Nineteen Peasant Wives*, trans. Taeko Midorikawa. New York: St. Martin's Press.

Kane, Kathleen O. 1994. "Hidden in Plain Sight: The Metaphysics of Gender and Death." Ph.D. diss., Department of Political Science, University of Hawai'i.

Kann, Mark. 1991. *On the Man Question: Gender and Civic Virtue in America*, Philadelphia: Temple University Press.

Keahiolalo, RaeDeen M. 1995. "A Site for Conquest—The Female Body in

Militarized Prostitution." Unpublished paper, Department of Political Science, University of Hawai'i, Honolulu, Hawai'i.

Kealoha-Scullion, Kehau. 1995. "The Hawaiian Journey: Out of the Cave of Identity, Images of the Hawaiian and Other." Ph.D. diss., Department of Political Science, University of Hawai'i.

Keir, Gerry. 1993. "A 'Last Roll Call' for the 442nd," *Honolulu Star-Bulletin*: A-8. March 27.

Kelly, Marion, and Nancy Aleck. 1997. *Mākua Means Parent*. Honolulu: American Friends Service Committee Hawai'i.

Kent, Noel. 1983. *Hawai'i: Islands under the Influence*. New York: Monthly Review Press.

Kilgore, M. L. n.d. "How Much Butter from How Many Guns?" Unpublished paper, Department of Political Science, University of Hawai'i.

Kiste, Robert C. 1974. *The Bikinians: A Study in Forced Migration*. The Kiste and Ogan Social Change Series in Anthropology. Menlo Park, Calif.: Cummings Publishing Company.

Klein, Bradley S. 1989. "The Textual Strategies of the Military; Or, Have You Read Any Good Defense Manuals Lately?" in *International/Intertextual Relations: Postmodern Readings of World Politics*, ed. James Der Derian and Michael J. Shapiro. Lexington, Mass.: Lexington Books, 97–110.

Knebel, Fletcher, ed. 1991. *The State of Journalism in Hawaii*. Honolulu: Honolulu Community-Media Council.

Kresnak, William. 1996. "Inouye Says Jobs Are Available," *Honolulu Advertiser*: A-1, A-2. February 24.

Kristof, Nicholas D. 1996. "Guam, a Pacific Paradise, Is Angry over Its Status as a Colony," *New York Times*: 14. November 10.

———. 1997. "An Atomic Age Eden (But Don't Eat the Coconuts)," *New York Times*: A-6. March 5.

Kubo, Louise. 1997. "Reading and Writing Local: A Politics of Community." Ph.D. diss., Political Science Department, University of Hawai'i.

Kuykendall, Ralph S. 1957. *The Hawaiian Kingdom 1778–1854: Foundation and Transformation*. Vol. 1. Honolulu: University of Hawaii Press.

———. 1966. *The Hawaiian Kingdom 1854–1874: Twenty Critical Years*. Vol. 2. Honolulu: University of Hawaii Press.

———. 1967. *The Hawaiian Kingdom 1874–1893: The Kalakaua Dynasty*. Vol. 3. Honolulu: University of Hawaii Press.

Landers, Peter. 1995. "Macke Storm Envelops Gore," *Honolulu Advertiser*: A3. November 19.

Layoun, Mary. 1992. "Telling Spaces: Palestinian Women and the Engendering of National Narratives," in *Nationalisms and Sexualities*, ed. Andrew Parker, Mary Russo, Doris Summer, and Patricia Yeager. New York: Routledge, 407–23.

Lee, Karen W. F. 1993. *Impact Aid and the Establishment of United States Department of Defense Schools in Hawaii.* Report #4. Honolulu: Legislative Reference Bureau.

Lefebvre, Henri. 1994. *The Production of Space,* trans. Donald Nicholson-Smith. Oxford: Basil Blackwell.

Leibowitz, Arnold. 1989. *Defining Status: A Comprehensive Analysis of United States Territorial Relations.* Boston: Martinus Nijhoff.

Lenihan, Daniel J., ed. 1989. *Submerged Cultural Resources Study: USS Arizona Memorial and Pearl Harbor National Historic Landmark.* Southwest Cultural Resources Center Professional Papers No. 23. Santa Fe: National Park Service, Submerged Cultural Resources Unit.

Liggett, Helen. 1994. "Urban Transgressions: Siting Memories and Dreams." Unpublished paper, Department of Urban Studies, Cleveland State University.

Lind, Andrew. 1968. "Service-Civilian Tension in Honolulu," in *Community Forces in Hawaii,* ed. Bernard Hormann. Honolulu: University of Hawaii Press, 248–53.

Lind, Ian. 1984/85. "Ring of Steel: Notes on the Militarization of Hawaii," *Social Process in Hawaii* 31: 25–47.

———. 1992. "Architects, Engineers Provide Bulk of Democratic Funds," *Hawai'i Monitor* 3 (1): 1–2. December.

Linnekin, Jocelyn. 1990. *Sacred Queens and Women of Consequence: Rank, Gender, and Colonialism in the Hawaiian Islands.* Ann Arbor: University of Michigan Press.

Little, Kenneth. 1991. "On Safari: The Visual Politics of a Tourist Representation," in *The Varieties of Sensory Experience: A Sourcebook in the Anthropology of the Senses,* ed. David Howes. Toronto: University of Toronto Press, 148–63.

Los Angeles Times. May 31, 1996a: A-1.

———. 1996b. June 8: A-1.

Lum, Darrell. 1990. "On My Honor," in Darrell Lum, *Pass On, No Pass Back!* Honolulu: Bamboo Ridge Press, 81–95.

Lutz, Catherine, ed. 1984. *Micronesia as a Strategic Colony: The Impact of U.S. Policy on Micronesian Health and Culture.* Cambridge, Mass.: Cultural Survival.

Lutz, Catherine, and Lesley Bartlett. 1995. *Making Soldiers in the Schools: An Analysis of the Army JROTC Curriculum.* American Friends Service Committee Report #18.

MacCaughey, Vaughan. 1916. "The Punchbowl: Honolulu's Urban Volcano," *Scientific Monthly:* 607–13. June.

MacKenzie, Melody Kapilialoha, ed. 1991. *Native Hawaiian Rights Handbook.* Honolulu: Native Hawaiian Legal Corporation.

Mackie, Vera. 1988. "Feminist Politics in Japan," *New Left Review* 167: 53–76. January/February.

Macleod, David I. 1983. *Building Character in the American Boy: The Boy Scouts, YMCA, and Their Forerunners, 1870–1920*. Madison: University of Wisconsin Press.

Mahan, Alfred Thayer. 1918. *On Naval Warfare*, ed. Allan Westcott. Boston: Little, Brown.

Marine Barracks of Hawaii Monthly Newsletter. 1990. 14 (November/December). Pearl Harbor: Puller Press.

Matsumoto, Alan. 1996. "Residents Endorse Military Presence," *Honolulu Star-Bulletin*: A-1, A-4. October 28.

McCartney, James. 1990. "B-2 Illustrates Folly of Secrecy," *Honolulu Star-Bulletin and Advertiser*: B-2. July 29.

McClintock, Anne. 1995. *Imperial Leather: Race, Gender, and Sexuality in the Colonial Contest*. New York: Routledge.

McGrath, Edward J. Jr., Kenneth B. Brewer, and Bob Krauss. 1973. *Historic Waianae: A Place of Kings*. Norfolk Island, Australia: Island Heritage.

McGregor, Davianna Pōmaikaʻi. 1997. "There's More to History," *Honolulu Advertiser*: B-1, B-3. September 7.

McHenry, Donald F. 1975. *Micronesia. Trust Betrayed: Altriusm vs. Self Interest in American Foreign Policy*. New York: Carnegie Endowment for International Peace.

Meinecke, Kalani. 1984. "Aloha ʻĀina," in *Hoʻihoʻihou: A Tribute to George Helm and Kimo Mitchell*, ed. Rodney Morales. Honolulu: Bamboo Ridge Press.

Memorandum to Program Directors and Administrators. 1992. Child and Family Services, State of Hawaii. November 19.

Menton, Linda, and Eileen Tamura. 1989. *A History of Hawaiʻi*. Honolulu: University of Hawaiʻi Press.

Merrill, Christopher. 1994. "A Little Justice in Hawaiʻi," *Nation* 259: 236. September 5/12.

Miller, Susan E. 1990. Statement on Behalf of the Natural Resources Defense Council before the Senate Committee on Governmental Affairs Regarding the Status of Federal Sites in Hawaii in Compliance with Federal and State Hazardous Waste Laws and Regulations. August 20.

Milner, Neil. 1992. "Single-Party Dominance Helps Inouye Prevail over Sex Charges," *Honolulu Star-Bulletin*: A-19. December 31.

Mitchell, Timothy. 1991. *Colonizing Egypt*. Berkeley and Los Angeles: University of California Press.

Mitford, Jessica. 1978. *The American Way of Death*. New York: Simon and Schuster.

Miyake, Yoshiko. 1991. "Doubling Expectations: Motherhood and Women's

Factory Work under State Management in Japan in the 1930's and 1940's." In *Recreating Japanese Women, 1600–1945*, ed. Gail Lee Bernstein. Berkeley and Los Angeles: University of California Press, 267–95.

Mock, Freida Lee, director. "Maya Lin: A Strong Clear Voice." 1995. Videotape. Santa Monica, Calif.: Sanders and Mock Productions.

Moffat, Riley M., and Gary L. Fitzpatrick. 1995. *Surveying the Mahele: Mapping the Hawaiian Land Revolution*. Vol. 2. Honolulu: Editions Limited.

Moi, Toril. 1985. *Sexual/Textual Politics: Feminist Literary Theory*. New York: Methuen.

Moore, Mike. 1995. "More Security for Less Money," *Bulletin of the Atomic Scientists*: 34–37. September/October.

Morrison, Toni. 1990. *Playing in the Dark: Whiteness and the Literary Imagination*. New York: Vintage Books.

Morse, Harold. 1997. "Waianae against Troop Landing at Makua," *Honolulu Star-Bulletin*: A-3. August 6.

Mosse, George L. 1990. *Fallen Soldiers: Reshaping the Memory of the World Wars*. Oxford: Oxford University Press.

Nader, Ralph, and Robert Fellmeth. 1972. "Daniel K. Inouye: Democratic Senator from Hawaii," in *Citizens Look at Congress*. Ralph Nader Congress Project. Written by Karen Winkler.

Nakamura, Barry S. 1979. "The Story of Waikīkī and the 'Reclamation' Project." Master's thesis, Department of History, University of Hawai'i.

Napier, A. Kam. 1995. "What's It All For? The Military in Hawai'i," *Honolulu Magazine*: 42–45, 114–21. August.

National Commission for Economic Conversion and Disarmament. 1995. "The Pentagon Tax Burden: Most States are Losers." Washington, D.C. April 13.

Nero, Karen L. 1989. "Time of Famine, Time of Transformation: Hell in the Pacific, Palau," in *The Pacific Theater: Island Representations of World War II*, ed. Geoffrey M. White and Lamont Lindstrom. Honolulu: University of Hawaii Press, 117–47.

Nielsen, Vicki. 1996. "Female Adolescents in the Military." Unpublished paper, Department of Social Work, University of Hawai'i, Honolulu.

Nietzsche, Friedrich. 1956a. *The Birth of Tragedy and the Genealogy of Morals*, trans. Francis Golffing. New York: Doubleday Anchor.

———. 1956b. "On Truth and Falsity in their Ultramoral Sense," in *The Complete Works of Friedrich Nietzsche*. Vol. 2: *Early Greek Philosophy*, ed. Oscar Levy and trans. Maximilian A. Mugge. New York: Doubleday Anchor, 175–87.

Nogelmeier, Puakea. 1985. *Puowaina*. Honolulu: Research and Development, Hawaiian Studies Institute, Kamehameha Schools/Bernice Pauahi Bishop Estate.

Nora, Pierre. 1989. "Between Memory and History: *Les Lieux de Memoire*," *Representations* 26 (Spring): 7–25.

O'Donnell, Shauna Maile. 1996a. "*Ke Ea O Ka 'Aina*." Unpublished paper, Environmental Studies Program, York University, North York, Ontario, Canada.

———. 1996b. "A 'New Choreography of Sexual Difference'; Or, Just the Same Old Song and Dance?" *Socialist Review* 25 (1): 95–118.

Ohira, Rod. 1996. "New Stealth Bomber Runs into Flak," *Honolulu Star-Bulletin*: A-3. May 28.

Okihiro, Gary T. 1991. *Cane Fires: The Anti-Japanese Movement in Hawaii, 1865–1945*. Philadelphia: Temple University Press.

O'Malley, Anne E. 1990. "Hawaiian Activist Says: It's Totally Stupid to Do It Here," *Weekend Kauai Times*: C-1, C-7. September 14.

Omandam, Pat. 1995a. "Inouye: EWC Must Accept Security, Military Mission," *Honolulu Star-Bulletin*: A-10. April 21.

———. 1995b. "In Search of a Nation," *Honolulu Star-Bulletin*: A-1. August 24.

———. 1997. "Makua Beach Is Spared—For Now," *Honolulu Star-Bulletin*: A-1. August 30.

Pacific War Memorial Commission. 1962. *Report . . . to the honorable members of the 1962 Hawaii Legislature*. Honolulu.

Pateman, Carole. 1988. *The Sexual Contract*. Palo Alto, Calif.: Stanford University Press.

Patrinos, Christine Diane. 1995. "Resistance Narratives in Hawai'i, 1893–1898." Ph.D. diss., Department of Political Science, University of Hawai'i.

Pennybacker, Mindy. 1996. "Should the Aloha State Say Goodbye? Natives Wonder," *Nation* 263 (5): 21–24. August 12/19.

Peterson, Spike, ed. 1992. *Gendered States: Feminist (Re)Visions of International Relations*. Boulder, Colo.: Lynne Rienner.

Pfund, Rose, ed. 1992. *Oil Spills at Sea: Potential Impacts on Hawaii*. Honolulu: University of Hawai'i Sea Grant Program.

Pichasie, Pete. 1996. "Inouye: Master of the Pork Barrel," *Honolulu Star-Bulletin*: A-1. March 6.

The Planter's Monthly. 1882. Vol. 1, March 22.

Politics Now. 1995. Online publication of the National Journal. http://www.politicsusa.com/PoliticsUSA/resources/almanac/his1.html.cgi

Porteus, Stanley D., and Marjorie E. Babcock. 1926. *Temperament and Race*. Boston: Richard G. Badger.

Pratt, Mary Louise. 1992. *Imperial Eyes: Travel Writing and Transculturation*. New York: Routledge.

Pratt, Minnie Bruce. 1984. "Identity: Skin, Blood, Heart," in *Yours in Struggle: Three Feminist Perspectives on Anti-Semitism and Racism* ed. Elly Bulkin, Minnie Bruce Pratt and Barbara Smith. New York: Firebrand Books: 11–63.

The Promised Land. 1819. Boston: Samuel T. Armstrong.

Public Affairs Office, Commander in Chief, U.S. Pacific Command. 1995. *Directory of Hawaii PAO/Media/PR Representatives.* Camp Smith, Hawaii.

Purnell, Nanette Napoleon. 1987. *Cemetery Research Project, 1987: Final Report.* Honolulu: Hawaiian Historical Society.

Rademaker, John Adrian. 1951. *These Are Americans.* Palo Alto, Calif.: Pacific Books.

Ramsey, Winston G., ed. 1982. *Pearl Harbor Then and Now.* After the Battle Series #38. London: Plaistow Press.

Rees, Robert. 1995. "Private Property vs. Native Hawaiian Rights." Videocassette, University of Hawai'i of Mānoa Library.

———. 1996. "Liar, Liar, Pants on Fire," *Honolulu Weekly* 6 (30): 4–6. July 24.

Rho, Marguerite. 1989. "Army Museum: The Long and Proud Military History of Hawaii Comes to Life within a Once-Mighty Defensive Fortress in Waikiki." *Ampersand:* 6–12. Summer.

Richardson, James O. 1973. *On the Treadmill to Pearl Harbor: The Memoirs of Admiral J. O. Richardson.* Washington, D.C.: Naval History Division.

Riveira, Berna-Lee Lehua, et al. 1990–91. "Hawaiian Women and the Struggle for Self-Determination of a People," *Voices* 4 (2): 6–17.

Robin, Ron. 1995. "'A Foothold in Europe': The Aesthetics and Politics of American War Cemeteries in Western Europe." *Journal of American Studies* 29(1): 55–72.

Rodriggs, Lawrence Reginald. 1991. *We Remember Pearl Harbor: Honolulu Civilians Recall the War Years 1941–1945.* Newport, Calif.: Communications Concepts.

Roelofs, H. Mark. 1992. *The Poverty of American Politics: A Theoretical Interpretation.* Philadelphia: Temple University Press.

Roff, Sue Rabbitt. 1991. *Overreaching in Paradise: United States Policy in Palau since 1945.* Juneau, Alaska: Denali Press.

Rohrer, Judy. 1994. "JROTC Expansion: The Defense Department Plan for Public Education," *Z Magazine* 37 (6): 35–39. June.

———. 1995. Appendix to "Hawaiian Sovereignty and the Military Occupation of Hawaiian Lands." Unpublished paper, Department of Political Science, University of Hawai'i.

————. 1997. "Haole Girl: Identity and White Privilege in Hawai'i," *Social Process in Hawai'i* 38: 138–61.

Rohter, Ira. 1992. *Green Hawaii: A Sourcebook for Development Alternatives*. Honolulu: Na Kane O Ka Malo Press.

Roosevelt High School JROTC. Home page. http://student\www.eng.hawaii.edu/yogi/RHS/rules.html

Rosenthal, Michael. 1984. *The Character Factory: Baden-Powell and the Origins of the Boy Scout Movement*. New York: Pantheon Books.

Roylance, Frank D. 1997. "Island Nations Raising Alarms on Rising Oceans," *Honolulu Advertiser*: A-24. September 21.

Sahlins, Marshall. 1995. *How "Natives" Think: About Captain Cook, for Example*. Chicago: University of Chicago Press.

Said, Edward W. 1979. *Orientalism*. New York: Vintage Books.

Scheer, Robert. 1997. "Our Rained-Out Bomber," *Nation* 265 (8): 6–7.

Schmitt, Robert C. 1995. "Rotten Statistics," *Honolulu Weekly* 5 (31): 7. July 26.

Seager, Joni. 1993. *Earth Follies: Coming to Feminist Terms with the Global Environmental Crisis*. New York: Routledge.

Seltzer, Mark. 1992. *Bodies and Machines*. New York: Routledge.

Sennott, Charles M. 1996. "Armed for Profit: The Selling of U.S. Weapons," *Boston Globe*: B-1–B-12. February 11.

Seto, Benjamin. 1990a. "Shipyard Wants Peace Dividend," *Honolulu Star-Bulletin*: A-1, A-4. August 14.

————. 1990b. "USS Missouri's Impact Divides Island Leaders," *Honolulu Star-Bulletin*: A-1, A-8. August 15.

Shapiro, Michael. 1997. "Sidewalk Sale," *Honolulu Weekly* 5–7. January 29–February 4.

Shapiro, Michael J. 1988. *The Politics of Representation: Writing Practices in Biography, Photography and Policy Analysis*. Madison: University of Wisconsin Press.

————. 1991. "Sovereignty and Exchange in the Orders of Modernity." Presented at the International Studies Association Convention, Vancouver, British Columbia. March.

————. 1993. *Reading "Adam Smith": Desire, History, and Value*. Newbury Park, Calif.: Sage Publications.

————. 1994. "Moral Geographies and the Ethics of Post-Sovereignty," *Public Culture* 6: 479–502.

————. 1997. *Violent Cartographies: Mapping Cultures of War*. Minneapolis: University of Minnesota Press.

Shapiro, Michael J., and Hayward R. Alker, eds. 1996. *Challenging Bound-*

aries: Global Flows, Territorial Identities. Minneapolis: University of Minnesota Press.

Sherry, Michael S. 1995. *In the Shadow of War: The United States since the 1930s.* New Haven, Conn.: Yale University Press.

Sibley, Mulford. 1970. *Political Ideas and Ideologies.* New York: Harper and Row.

Silva, Noenoe. 1997. "Kūʻē! Hawaiian Women's Resistance to the Annexation," *Social Process in Hawaiʻi* 38: 2–15.

———. Forthcoming. *"Ke Kūʻē Kūpaʻa loa nei Mākou: A Study of Hawaiian Resistance to Colonization."* Ph.D. diss., Department of Political Science, University of Hawaiʻi.

Simpson, MacKinnon. 1993. *The Lymans of Hawaiʻi Island: A Pioneering Family.* Hilo, Hawaiʻi: Orlando H. Lyman Trust.

Singletary, Millie. 1977. *Punchbowl: National Memorial Cemetery of the Pacific.* Honolulu: MFS.

Sinha, Mrinalini. 1995. *Colonial Masculinity: The 'Manly Englishman' and the 'Effeminate Bengali' in the Late Nineteenth Century.* Manchester, England: Manchester University Press.

Smith, Stanley H., comp. 1990. *Investigations of the Attack on Pearl Harbor: Index to Government Hearings.* Westport, Conn.: Greenwood Press.

Smyser, A. A. 1991. "Hawaii Has Come a Long Way Since 1945," *Honolulu Star-Bulletin*: A-10. January 1.

———. 1995. "Inouye's Talk at the East West Center," *Honolulu Star-Bulletin*: A-12. April 25.

Soguk, Nevzat. Forthcoming. *States and Strangers: Refugees and Displacements of Statecraft.* Minneapolis: University of Minnesota Press.

Spain, Lauren. 1995. "Cut Foreign Arms Sales," *Bulletin of the Atomic Scientists* 51 (5): 47–48. September/October.

Sprague, Roberta A. 1988. "Wilkes Expedition—Framework for American Expansionism: The United States Exploring Expedition, 1838–1842." Master's thesis, Department of History, University of Hawaiʻi.

Stannard, David E. 1989. *Before the Horror: The Population of Hawaiʻi on the Eve of Western Contact.* Honolulu: Social Science Research Institute.

Stiehm, Judith H. 1983. "The Protected, the Protector, the Defender." In *Women and Mens' Wars*, ed. Judith H. Stiehm. Oxford: Pergamon Press, 367–76.

Sturken, Marita. 1991. "The Wall, the Screen and the Image," *Representations* 35: 118–42. Summer.

Sullam, Brian. 1978. "Dan Inouye: The Uses of Power," *Hawaii Observer* 22. February 23.

Sylvester, Christine. 1994. *Feminist Theory and International Relations in a Postmodern Era*. Cambridge: Cambridge University Press.

Tamura, Eileen. 1994. *Americanization, Acculturation and Ethnic Identity: The Nisei Generation in Hawaii*. Urbana: University of Illinois Press.

Tanahara, Kris, and Jon Yoshishige. 1992. "Wife Declares Her Trust in Inouye," *Honolulu Advertiser*: A-1. October 18.

Tavares, Nils. 1945. "Prostitution and the Law." Presented at the Meeting of the Social Protection Committee of the Honolulu Council of Social Agencies. February 7.

TenBruggencate, Jan. 1991. "Inouye to Visit Gulf at Cheney's Request," *Honolulu Advertiser*: A-3. February 14.

———. 1997. "Seabee Bridge Bypasses Kauai Red Tape," *Honolulu Advertiser*: A-1, A-17. July 3.

Thurmond, Strom. 1995. "Revitalization of National Defense," *The Officer*: 23–24, 34. March.

Tickner, J. Ann. 1992. *Gender in International Relations: Feminist Perspectives on Achieving Global Security*. New York: Columbia University Press.

Todorov, Tzvetan. 1984. *The Conquest of America: The Question of the Other*, trans. Richard Howard. New York: Harper and Row.

Transactions of Royal Hawaiian Agricultural Society Including a Record of the Proceedings to the Formation of the Society. 1850. 1 (1). Honolulu: Henry M. Whitney Government Press.

Transactions of the Royal Hawaiian Agricultural Society. 1851. J. F. B. Marshall, chairman. Report of the Committee on Labor.

———. 1853. J. F. B. Marshall, manager of Lihue Plantation to Geo. E. Lathrop, M.D., chairman. Report of the Committee on Labor.

Trask, Haunani-Kay. 1993. *From a Native Daughter: Colonialism and Sovereignty in Hawai'i*. Monroe, Me: Common Courage Press.

Tummons, Patricia. 1991a. "Bay of Infamy." *Honolulu Weekly* 1 (21): 1, 4–5, 14. December 4.

———. 1991b. "Army Displays Its Brass at Schofield Barracks," *Environment Hawai'i* 2 (4): 5–6, 8. October.

———. 1991c. "The Navy's Superfund Six," *Environment Hawai'i* 2 (6): 3–4. December.

———. 1991d. "Oil Contamination Is Pervasive," *Environment Hawai'i* 2 (6): 5–7. December.

———. 1991e. "Remember Pearl Harbor: A Call to Arms for Environmentalists," *Environment Hawai'i* 2(6): 1, 7–8. December.

———. 1992a. "Waikane War Zone," *Honolulu Weekly* 2 (40): 1, 4–5, 7. September.

————. 1992b. "From Fertile Fields to No-Man's Land: The Transformation of Waikane Valley." *Environment Hawai'i* 3 (2): 1, 7–9. August.

————. 1992c. "Use of Islands by Armed Forces Leaves Few Stones Unturned," *Environment Hawai'i* 3 (2): 4–7. August.

————. 1993a. "Clamming Up at Pearl Harbor," *Environment Hawai'i* 4 (3): 1. September.

————. 1993b. "Spent Submarine Reactor Fuel to Be Kept at Pearl Harbor 'til 1995," *Environment Hawai'i* 4 (3): 1, 6. September.

————. 1993c. "When It Comes to Waste, Navy Has It All," *Environment Hawai'i* 4 (3): 4–5. September.

————. 1994. "Marine Projects Designed to Help Stilts, Boobies," *Environment Hawai'i* 5 (3): 10. October.

Twigg-Smith, Thurston. 1997. "Sovereignty Advocates Are Wrong about History," *Honolulu Advertiser*: B-1, B-3. September 7.

United States Army ROTC. 1989. *Leadership Education and Training.* Vol. 1. Ft. Monroe, Va.: U.S. Army ROTC Cadet Command.

United States Congress. 1976. *Report of the Joint Committee on the Investigation of the Pearl Harbor Attack. Including the Minority Report.* 79th Cong., 2nd Sess. Reprint. New York: Da Capo Press.

United States Congress. 1993. *Hearings before a Subcommittee of the Committee on Appropriations United States Senate.* "Pacific Rim Issues." 103rd Cong. Washington, D.C.: Government Printing Office.

United States Bureau of Education. 1920. "A Survey of Education in Hawaii." Washington, D.C.: Government Printing Office.

United States Department of Defense. 1990. *Oahu Telephone System.* Hawaii: Military Oahu Telephone System.

————. 1995. *Hawaii Land Use Master Plan/United States Pacific Command.* Camp H. M. Smith, Hawaii: U.S. Pacific Command.

USASCH (United States Army Support Command, Hawaii). 1979. *Environmental Impact Statement.* Fort Shafter, Hawaii.

Vartabedian, Ralph. 1997. "Lockheed, Northrop to Merge: $1.6 Billion Deal May Cost 10,000 Jobs," *Honolulu Advertiser*: B-8, B-7. July 4.

Vlahos, Michael. 1995. "Are We Prepared to Meet the Threat?" *The Officer*: 33, 41. August.

Waite, David. 1990. "Ordnance Called Big Hazard Here," *Honolulu Advertiser*: A-3. August 21.

Webb, Gary. 1996. "The Dark Alliance," *San Jose Mercury News*, August 18, 19, 20.

Weisgall, Jonathan. 1994. *Operation Crossroads: The Atomic Tests at Bikini Atoll.* Annapolis, Md.: Naval Institute Press.

————. 1996. "Fifty Years of the Atomic Age: Bikini Atoll, the Pacific and

Today's Students," presented at the University of Hawai'i at Manoa. October 16.

Wheeler, Gerald E. 1963. *Prelude to Pearl Harbor: The United States Navy and the Far East, 1921–31*. Columbia: University of Missouri Press.

Whittaker, Elvi. 1986. *The Mainland Haole: The White Experience in Hawaii*. New York: Columbia University Press.

White, Geoffrey M. 1998. "Moving History: The Pearl Harbor Film(s)," *positions* 5(3): 709–44.

White, Geoffrey M., and Lamont Lindstrom, eds. 1989. *The Pacific Theater: Island Representations of World War II*. Honolulu: University of Hawai'i Press.

White, Hayden. 1978. *Tropics of Discourse: Essays in Cultural Criticism*. Baltimore: Johns Hopkins University Press.

Whitney, Scott. 1995. "The War with the Ancestors," *Honolulu*: 46–48, 110–12. August.

Wilkes, Charles. 1970. *United States Exploring Expedition*. Vols. 1–5. Ridgewood, N.J.: Gregg Press. Originally printed in Philadelphia: Lea and Blanchard, 1845.

Wilson, Lynn B. 1995. *Speaking to Power: Gender and Politics in the Western Pacific*. New York: Routledge.

Witty, Jim. 1997. "Army Open House at Makua to Try to Win over Public," *Honolulu Star-Bulletin*: A-1, A-8. August 4.

Wright, Theon. 1966. *Rape in Paradise*. New York: Hawthorne Books.

Wright, Walter. 1996a. "Protestors Disrupt B-2 Naming Rite," *Honolulu Advertiser*: A-1. May 28.

———. 1996b. "After 50 Years, Bikini Islands Still Waiting," *Honolulu Advertiser*: B-1. October 15.

Yamaguchi, Andy. 1992. "Inouye Honored by Nurses," *Honolulu Advertiser*: A-3. October 23.

Yoshishige, Jon. 1991. "Inouye: Don't Forget Viet Vets," *Honolulu Advertiser*: C-6. May 27.

———. 1993. "Fuel Tanks atop Primary Water Source," *Honolulu Advertiser*: A1, A-19. October 17.

Zoll, Daniel. 1998. "Weapons of Mass Distraction," *Honolulu Weekly* 8(14): 6–8. April 8–14.

Interviews

Abercrombie, Neil. December 15, 1993. Congressman, U.S. House of Representatives, First District, Hawai'i.

Ali, Mehmed. January 15, 1991. Corporal, U.S. Marines.

Barrell, Lt. Col. Donald, and Lt. Col. Jeff Tom. July 10, 1995. Barrell, now retired, was statewide director for JROTC for the Department of Education, a position that his deputy, in 1995, now holds.

Bowman, Judith. July 24, 1997. Curator, Fort DeRussy Army Museum.

Burnette, Christine. March 12, 1991. Director of Family Support Services, Hickam Air Force Base.

Fujii, Dennis. March 12, 1991. Chief of Community Relations, Hickam Air Force Base.

Harris, George. March 15, 1991. Lieutenant, U.S. Navy.

Imai, Eugene. November 22, 1995. Vice President for Administration, University of Hawai'i, and General, Hawaii Army National Guard.

Inouye, Daniel. October 23, 1996. U.S. Senator, Hawai'i.

Johnson, Timothy M. January 31, 1991. Captain, Army ROTC, University of Hawai'i. Recruiting Operations Officer.

Keefer, Joel. April, 1991. Commander, U.S. Navy. Pearl Harbor.

McCouch, Nelson. March 8, 1991. U.S. Army Captain, Public Affairs Officer. 25th Infantry Division, Schofield Barracks.

Mitchell, Sam. July 21, 1997. President, Machinists and Aerospace Workers Local 1998, Pearl Harbor Shipyard.

Pfister, Jan Timothy. May 5, 1991. United States Marine Corps, CINCPACFLT. Camp H. M. Smith.

Index

KATHY E. FERGUSON is professor of political science and women's studies at the University of Hawai'i at Mānoa.

PHYLLIS TURNBULL is associate professor of political science at the University of Hawai'i at Mānoa.

Printed and bound by CPI Group (UK) Ltd, Croydon, CR0 4YY

09/10/2024

14571398-0002